Space 2030

TACKLING SOCIETY'S CHALLENGES

ORGANISATION FOR ECONOMIC CO-OPERATION AND DEVELOPMENT

ORGANISATION FOR ECONOMIC CO-OPERATION AND DEVELOPMENT

The OECD is a unique forum where the governments of 30 democracies work together to address the economic, social and environmental challenges of globalisation. The OECD is also at the forefront of efforts to understand and to help governments respond to new developments and concerns, such as corporate governance, the information economy and the challenges of an ageing population. The Organisation provides a setting where governments can compare policy experiences, seek answers to common problems, identify good practice and work to co-ordinate domestic and international policies.

The OECD member countries are: Australia, Austria, Belgium, Canada, the Czech Republic, Denmark, Finland, France, Germany, Greece, Hungary, Iceland, Ireland, Italy, Japan, Korea, Luxembourg, Mexico, the Netherlands, New Zealand, Norway, Poland, Portugal, the Slovak Republic, Spain, Sweden, Switzerland, Turkey, the United Kingdom and the United States. The Commission of the European Communities takes part in the work of the OECD.

OECD Publishing disseminates widely the results of the Organisation's statistics gathering and research on economic, social and environmental issues, as well as the conventions, guidelines and standards agreed by its members.

> *This work is published on the responsibility of the Secretary-General of the OECD. The opinions expressed and arguments employed herein do not necessarily reflect the official views of the Organisation or of the governments of its member countries.*

Also available in French under the title:
L'espace à l'horizon 2030
RELEVER LES DÉFIS DE LA SOCIÉTÉ DE DEMAIN

© OECD 2005

No reproduction, copy, transmission or translation of this publication may be made without written permission. Applications should be sent to OECD Publishing: *rights@oecd.org* or by fax (33 1) 45 24 13 91. Permission to photocopy a portion of this work should be addressed to the Centre français d'exploitation du droit de copie, 20, rue des Grands-Augustins, 75006 Paris, France (*contact@cfcopies.com*).

Foreword

This publication is the final report on a two-year OECD Futures Project devoted to the commercialisation of space, the development of space infrastructure and the role to be played by public and private actors (the Space Project).

The project was carried out over 2003-04 and involved 25 public and private participants from the OECD area. It was launched in response to growing concern in many quarters that the potential contribution of space to the economy and society at large was not being fully exploited and that the topic would greatly benefit from policy attention beyond the space community itself. It was felt that to support the policy formulation process in OECD countries and beyond, there was a need for a forward-looking, impartial analysis of the role space might play in future and of the actions needed to ensure that space contributes fully to economic and social development. It was further thought that, to ensure genuine impartiality and a fresh perspective, the analysis should be conducted by a "non-space" actor, i.e. by an organisation not traditionally associated with the space sector itself.

Several factors made the OECD a particularly appropriate platform for this project. First, many OECD countries devote a significant amount of resources to space-related activities. Second, most of the key players, both public and private, are located in the OECD area. Third, space applications will increasingly concern domains where OECD governments have major interests beyond their traditional responsibilities for military and scientific fields (e.g. security, environment, education, health, communications and transport). Fourth, many of the policy issues raised by the future development of the space sector (e.g. regulation of markets, industrial and scientific policy, public and private governance) clearly fall within the OECD's field of competence. Finally, as a non-space actor, as defined above, the OECD can act as an honest broker and offer an impartial forum for addressing space-related issues.

Following extensive consultations in 2002 with leading public and private space actors, the OECD Secretariat formulated in October 2002 a project proposal that was used as the basic roadmap for the study. The project was conducted by a team in the OECD's International Futures Programme (IFP), a forward-looking, multidisciplinary unit with a mandate to alert the Secretary-General and the Organisation to emerging issues by pinpointing major developments and analysing key long-term concerns, with a view to helping governments map strategy.

The project provides a "non-space" assessment of the opportunities and challenges facing the space sector, particularly in terms of civil applications, and outlines what needs to be done to make its contribution to society at large as effective as possible.

FOREWORD

The analysis was broken down into five main phases:

- *In the first two phases of the project, the current state of the space sector was examined and its possible future evolution under alternative scenarios explored, with a view to identifying space applications that might be considered "promising" – from a societal perspective – over the long term.*

- *In the third phase of the project, an in-depth analysis of some of the applications identified in the second phase was conducted in order to achieve a more concrete appreciation of the factors – notably government action – that would be essential to their success, as well as of the contribution such applications could make to society at large.*

- *Building on this work and taking into account more generally the potential contribution of space to meeting future societal challenges, attention focused in the fourth phase on the framework conditions (including institutional, legal and regulatory conditions) that govern space activities.*

- *In the fifth and final phase of the project, the main conclusions drawn from the analysis were outlined, and recommendations were set out that offer a long-term, future-oriented framework for decision makers.*

Throughout, extensive consultations were conducted with project participants, including four meetings of the Project Steering Group. Moreover, special working groups were set up in the third phase, during which a more in-depth analysis of certain applications was conducted. The work also benefited from contributions by a number of experts who were called upon to address specific issues and to provide comments on drafts. Finally, very valuable input was received from a number of international organisations with an interest in space-related issues.

This publication presents the main conclusions reached during the five main phases of the project, although it focuses primarily on the last two (framework conditions and recommendations). A previous publication, Space 2030: Exploring the Future of Space Applications, issued in May 2004, covers the first two phases of the project. Expert reports prepared as background material to this work are available on the OECD Web site. A forthcoming publication will be devoted to the in-depth case studies conducted in the third phase of the project.

Michel Andrieu is the principal author of this publication. Claire Jolly contributed to key chapters of the report. Advice was provided by Pierre-Alain Schieb, the initiator and co-ordinator of the project, and by Barrie Stevens who is directing the preparation of the series of reports. Anita Gibson and Manon Picard provided technical support, and Doranne Leclercle the editing.

Michael Oborne
Director of the OECD International Futures Programme
February 2005

Acknowledgments

The Space Project has been a stimulating adventure for all members of the Project Team, not only because the issues involved are both fascinating and important from a societal perspective, but also because the various tasks accomplished in the course of the Project have given us the opportunity to establish, with numerous experts in a broad range of domains, a lively and constructive dialogue that has considerably enriched our own thinking on the subject.

The consultations we have undertaken have been far-ranging and it is not possible to acknowledge everyone here. However, we wish to express our sincere appreciation to those who have been particularly helpful in providing us with comments, feedbacks and advice.

First and foremost we would like to thank all the members of the Steering Committee (see Annex B). Without them, this project would not have been possible. Special thanks for their input go to Karl-Heinz Kreuzberg, Josef Aschbacher and Anders Jordhal of the European Space Agency, Timothy Striker and Mara Browne of NOAA, David Abelson and Bill Lynch of Lockheed Martin, Christine Leurquin and Jean Paul Hoffman of SES Global, Didier Vassaux of CNES, Paula Freedman of the British National Space Centre and to Jon Wakeling of British Telecom.

Many professionals in the space sector and in government user departments also contributed their views, through meetings and via e-mail, and we would like to thank them all. We are particularly grateful to Ray Harris (University College London, United Kingdom), Philippe Munier (Spot Image, France), Patrick Collins (Azabu University, Japan), David Ashford (Bristol SpacePlanes, United Kingdom), Lucien Deschamps (CNRS, France), Pierre Lionnet (Eurospace, France), Laurent Gathier (Dassault, France), Paul Heinerscheid (Satlynx, Luxembourg) and Didier Faivre (ESA, France).

We would also like to thank representatives of various international organisations who kindly contributed to our work, notably: Rodolfo de Guzman (World Meteorological Organisation) and Tomoko Miyamoto (World Intellectual Property Organisation).

A number of OECD colleagues also contributed, offered useful advice or commented in writing. In this regard we are particularly grateful to Dirk Pilat (STI) and to Sandrine Kergroach-Connan (STI/EAS) who contributed a very useful paper on measuring the space economy.

Last but not least, we would like to acknowledge the contribution of the experts who were commissioned to provide their advice and assistance in various phases of the project. Special thanks in this regard are extended to Henry Hertzfeld (Space Policy Institute, George Washington University, United States) who provided very useful input to several phases of the project. We would also like to thank Frans van der Dunk (Leiden Institute of Space Law) for his help in the last two phases of the project and Alain Dupas for his contribution to the second and third phases.

Annex C provides a complete listing of the experts who contributed to the project and to whom we are very grateful. All errors, of course, remain our own.

The Project Team

Table of Contents

Executive Summary 9

Introduction 19

Chapter 1. **A Scenario-based Exploration of the Future of Space Applications** 21

Chapter 2. **Meeting Societal Challenges: How Space Might Help** 53

Chapter 3. **Supply Conditions: Strengths and Weaknesses of Space** 91

Chapter 4. **Framework Conditions: Institutional Aspects** 129

Chapter 5. **Framework Conditions: Legal, Regulatory and Public Awareness Aspects** 171

Chapter 6. **Main Findings and Recommendations** 205

Annex A. **Case Studies on Selected Space-based Applications** 273
Annex B. **The Space Project Steering Group** 321
Annex C. **Experts Involved in the Project** 324

Acronyms 329

List of boxes

2.1. International initiatives for developing the use of satellite Earth observation data 57
2.2. Estimated benefits of forecasting El Niño using space data 59
2.3. The Kyoto Protocol 61
2.4. Major risks facing society in coming decades 75
2.5. Knowledge and economic development 80
2.6. The impact of Landsat 7 hardware failure on post fire assessment.. 86
3.1. Japan's R&D efforts in satellite broadband 100
3.2. Autonomous in-orbit rendezvous and docking: first steps for in-orbit servicing capabilities 105
3.3. International co operation in semiconductors through SEMATECH.. 110
3.4. The Missile Technology Control Regime (MTCR) 115
3.5. WiMax: a new disruptive technology? 117

4.1. The European technology road map harmonisation process 142
4.2. Measuring the performance of a navigation system. 155
5.1. United Nations treaties and main resolutions concerning
space activities 173
5.2. Sea Launch and the concept of launching state in international law 177
5.3. What is UNIDROIT? 179
5.4. Limiting the liability of commercial launch operators 182
5.5. The allocation of satellite orbital positions and frequency use 185
5.6. International bodies active in the standardisation of space systems . . . 187
5.7. The US regime for technology transfer: International Traffic
in Arms Regulations (ITAR) 192
5.8. The Inter-Agency Space Debris Co-ordination Committee 194

List of figures

2.1. Transport growth by mode in western Europe since 1965 67
2.2. Scheduled passenger traffic, 1983-2002 68
2.3. Personal transport activity by region, 2000-50 70
2.4. Road and rail freight transport activity by region, 2000-50 71
6.1. Concept of the recommendations: "Building a sustainable bridge
to the future for the governmental and private actors active
in the space sector" 212
A.1. Main actors involved in the provision of telehealth services 276
A.2. Main actors involved in the provision
of satellite entertainment services 283
A.3. The broadband entertainment value chain 284
A.4. General system mapping for LBS and road management 289
A.5. Risk and disaster management cycle 295
A.6. The systemic view for risk and disaster management. 296
A.7. System mapping for space tourism. 305
A.8. Cash flow generated by the space tourism company 307

Liste of tables

1.1. Broad categories of space applications 40
1.2. Potentially promising applications 47
1.3. The synthesis scenarios 48
4.1. Space institutions and lines of reporting. 138

ISBN 92-64-00832-2
Space 2030
Tackling Society's Challenges
© OECD 2005

Executive Summary

EXECUTIVE SUMMARY

A number of important studies have been carried out on the space sector in recent years by national governments, research institutions and international organisations. This publication builds on previous work but takes a somewhat different perspective: its broad socio-economic approach and geographical coverage set it apart from earlier reports.

Most past studies of the space sector have focused on the supply side: technological advances and the types of new capabilities that can be developed. They assume, often incorrectly, that development eventually follows such advances. This publication explores instead how governments can get the most out of future public and private space investment. It focuses on efforts to develop a space infrastructure prepared to respond fully to future challenges, as well as on actions that governments might take to foster the use of space, when it is cost-effective, and to create a more supportive business environment.

The publication explores first the long-term demand for space applications, using a scenario-based approach to consider the role that space – military, civil and commercial – might play in alternative visions of the future. This broad assessment of long-term demand is enriched by a more detailed analysis focused on the contribution such applications might make to resolving some of the major societal challenges to be faced in the coming decades. Next, the supply side strengths and weaknesses of space solutions for fulfilling their promise and the key role of framework conditions are assessed. This sets the stage for the formulation of recommendations that are intended to provide decision makers with a broad policy framework they can use in their own policy work.

A scenario-based exploration of the future demand for space applications

A scenario-based analysis, with a 30-year time frame, was constructed on the basis of the interaction of three main drivers of societal change: geopolitical, economic and environmental. The results (published in *Space 2030: Exploring the Future of Space Applications*, OECD, 2004) suggest high demand for space across a range of very different views of the future. This is partly because military space is likely to play an important role in all the scenarios. However, civil and commercial space were also found to have generally favourable prospects, primarily because they are expected to be increasingly needed to tackle many of the world's social, technical and environmental problems.

Meeting societal challenges: how space might help

What, then, are the key long-term challenges facing society in which space might play a significant, constructive role? Devoting resources to the development of space systems can only be justified from a socio-economic perspective if significant benefits accrue to society at large. To address this question, the publication identifies and explores five major challenges for which space might help, namely, challenges related to the environment, the use of natural resources, the increasing mobility of people and goods and its consequences, growing security threats and the move towards the information society.

The analysis leads to the conclusion that space applications have indeed substantial potential to help tackle these problems. However, for that potential to be realised, a number of important conditions must be met: space systems require further development, need to be better integrated with ground-based systems and be provided in a sustainable and predictable manner.

Supply conditions: strengths and weaknesses of space

Whether these conditions will actually be met is far from clear. There are two main sources of concern: first, the rather weak current state of the sector – as noted in the first phase of the project – and second, the need for space actors to overcome major technological and economic challenges in coming decades.

Today, the sector's upstream component is subject to chronic excess supply, its downstream component is unevenly developed, and both are highly cyclical. Moreover, space business is not business as usual. In particular, governments continue to play a dominant role. First, the public sector is the main actor in conducting basic R&D and in developing space systems; it also plays a key role in the operation of such systems. Second, government agencies are the largest buyers of space goods and services. Third, governments set the framework conditions that govern private space activities and intervene heavily in the activities of private-sector actors, often for strategic reasons, because of the dual civil/military nature of space technology.

Given this state of affairs, one may wonder whether space actors will be in a position to tackle effectively the serious technological and economic hurdles that will need to be overcome if space is to fulfil its potential. This will very much depend on whether the framework conditions (i.e. institutional arrangements as well as laws and regulations) that govern space activities provide a supportive environment for such activities. The case studies conducted in the third phase (see Annex A) confirm the importance of such framework conditions.

Framework conditions: institutional aspects

Broadly speaking, the space sector involves three sets of actors: space agencies, public and private operators of space applications, and the upstream segment of the industry (*e.g.* spacecraft and launcher manufacturers and providers of launching services).

From an institutional perspective, the main question is the role to be assigned to each actor and the relationships among them. The answer will largely depend on the importance decision makers give to space, on the specific priorities they assign to space activities (*e.g.* space exploration, basic R&D, development of space applications) and on their views on the roles of public and private actors. Moreover, institutional arrangements are not static; they evolve over time to take account of changing conditions (*e.g.* as technology matures or as the economic and geopolitical environment evolves).

Many of the questions to be addressed relate to the status of space agencies: What should be the main focus of their activities? What relations should be forged with other ministries, notably user ministries and the ministry of defence? How should international co-operation with other agencies be carried out?

Another set of issues relates to the operation of space applications and to whether they should be operated by public or private actors. The answer depends largely on the nature of the application, the size of the market and on public policy towards the private sector. Different countries may adopt different solutions, distorting, in some cases, competition at international level.

Framework conditions: legal, regulatory and public awareness aspects

The legal and regulatory framework plays a central role in shaping space activities because it determines the rules of the game under which space actors – notably private ones – operate.

The legal framework. Although a number of basic components of the legal framework are in place (international space law regime and legislation at national level in some countries), major gaps remain. First, a number of countries still do not have national space laws. This is a source of uncertainty for space actors, especially private ones. Second, because international space law is a public regime, it is not well-suited to business transactions. Third, existing national space laws are not always business-friendly, as they were often developed with a view to security and strategic considerations, rather than with business in mind.

The regulatory framework. This should ideally provide basic rules of the game that help establish a stable and predictable environment for business and also stimulate innovation and encourage entrepreneurship. This is far from the case. For instance, the International Telecommunications Union (ITU) procedure for the allocation of frequencies and orbital slots raises a number of issues and is a source of uncertainties. World Trade Organisation (WTO) discipline is limited in most space product and service markets. Export controls tend to restrict the ability to exploit market opportunities. The space debris issue is inadequately addressed, contributing to increased risks of failure and costs for operators of spacecraft. Several standardisation questions remain open. This state of affairs hampers competition and discourages innovation and investment in the development of space systems.

Public awareness. Another difficulty results from the lack of visibility of space activities among the general public, whose general perception tends to be distorted, as the media focus almost exclusively on sensational successes and failures. As a result, the general population has a poor understanding of the value of space-based services in their daily lives and therefore does not fully support space activities. Finally, few students are inclined to embrace space-related careers.

Main findings and recommendations

Several important conclusions follow from the analysis that was conducted in the various phases of the project.

Lessons learned during the project

The first conclusion is that the future demand for space applications is likely to be substantial. This finding results from the analysis presented in Chapters 1 and 2.

This somewhat rosy vision of the future is tempered by a second finding which suggests that severe short- and medium-term fluctuations are likely to affect space actors, given the capital-intensive nature of space activities, the long lead times required for the development of space assets, the high risks of space ventures and the heavy involvement of the state in space activities.

Moreover, it was found that a clear distinction needs to be made between the upstream segment of the sector (*i.e.* space asset manufacturing and launching services) and the downstream segment (space applications). Typically, the downstream segment offers better prospects over the longer term than the upstream segment, which suffers from a situation of chronic oversupply owing largely to a desire of governments of space-faring nations to establish and maintain, for strategic and national sovereignty reasons, independent access to space.

It was also noted that, while the downstream sector offers the best prospects overall, not all applications are equally promising over the 30-year period under consideration. Information-intensive applications such as satellite-based telecommunications, Earth observation and navigation have a bright future. On the other hand, the prospects of transport and manufacturing applications are more uncertain, given the expectation that the cost of access to space is unlikely to decline drastically over the period, and given the complex technical problems of working in space.

Critical success factors. An important result of the analysis conducted in the third phase of the project and summarised in Annex A is that that there are significant commonalities across applications regarding the factors and issues that are critical for their successful development. First, the importance of a stable and predictable environment was evident in all case studies. Another strong message is the need to deal effectively with uncertainties that relate to liability, notably for emerging applications, and the importance of creating and preserving a balanced competitive environment when the services provided by the applications under consideration have to compete with services offered by other actors.

Equitable access to services was another major theme, one which extends beyond the digital divide between rural and urban dwellers to encompass questions of equal treatment of individual and national entities regarding access to information and knowledge derived from space activities in general.

In most of the case studies, issues related to the generation, distribution and use of information also played a prominent role, notably questions related to intellectual property, the pricing of data and the problem of data confidentiality and privacy.

The case studies also demonstrated that greater compatibility of technological systems, standards, licensing practices and so on are central to the future development of space applications. Moreover, the key role of infrastructure and the extent to which public authorities should be involved in its provision and operation are recurring issues. Finally, in a number of instances, there was a clear-cut case to be made for encouraging government support of R&D.

Many of these critical issues fall squarely under the responsibility of governments. Moreover, they go far beyond the traditional field of space policy and should be considered in a much broader policy context (*e.g.* economic, social and environmental policies).

The analysis conducted in the fourth phase of the project complements and confirms the findings of previous phases. It clearly indicates that the potential of space will not be realised unless governments take decisive action to improve the framework conditions that govern space activities.

Purpose, scope and general architecture of the recommendations

Purpose. The recommendations drafted on the basis of these findings are intended to provide an overall, consistent set of broad policy orientations that can offer a useful framework for policy formulation. The recommendations take the perspective of society in general rather than that of the space sector. Hence, they address governments rather than the space community as such. More specifically, they are intended for ministries with responsibility for overall economic and social policies that may bear on the performance of private space actors, as well as for user departments that might take advantage of space-based solutions for delivering their services to the general public.

Scope. The recommendations focus on the "big picture" and take a long-term policy view. They concern what might be done to strengthen the contribution that space can make to solving important socio-economic challenges. Moreover, they go beyond the traditional ambit of space policy to policy areas that may have a bearing on the successful deployment and use of space applications to meet societal challenges. The actions are proposed for the short to medium term, even though the project has looked at the space sector from a long-term perspective throughout.

It is beyond the scope of this project to assess whether existing policies are consistent with these recommendations, although that might be a logical step for follow-up work. It would require a thorough review of existing policies in the various member countries. Moreover, the recommendations take a socio-economic perspective and should therefore be viewed as only one of the inputs in a comprehensive decision-making process, which should also take into account strategic and geopolitical considerations, for example.

General architecture of the recommendations. The recommendations are constructed with a "bridge" in mind: How do we get from "here and now" to a much-improved situation 20 years or so down the road? The "surface" of the bridge consists of three blocks of roughly equal importance that stand for a cluster of policies for achieving a specific but broad-based objective. Each is supported by a number of "pillars", each of which represents a policy thrust aimed at achieving the block's broad objective and is composed of a set of specific complementary recommendations. This approach to the formulation of the recommendations makes it possible to clearly establish the context of each recommendation, its relative importance and how it relates to others in the overall policy framework.

A common format is adopted in order to present the recommendations in a consistent and systematic manner. For each recommendation, the following structure is used:

- *General view* sets the context in which the recommendation is made.

- *Why* explains why government action is needed.
- *How* outlines actions governments might take to fulfil the recommendation.
- *Examples* shows how some of the actions identified have been implemented in some countries.

Because of the systematic approach adopted here, a number of the recommendations presented in this publication are not new. However, the merit of the approach is that it shows the relative importance of each and how it fits into an overall socio-economic strategy for the sector. Moreover, the term "special focus" identifies specific recommendations that may perhaps open interesting new perspectives.

The three main blocks of the recommendations are as follows:

- **Block 1: Implement a sustainable space infrastructure.** Without an efficient, robust and sustainable infrastructure, space actors will not be able to deliver space-based services that fully meet societal needs in an effective and efficient manner. This should be a major thrust of public policy, since governments play the dominant role in the development of infrastructure.
- **Block 2: Encourage public use.** Space infrastructure offers very attractive opportunities for pursuing a broad range of public missions in a cost-effective manner. From an overall public governance perspective, these opportunities need to be fully exploited by public actors, and the development of new and innovative services for the public at large that may result needs to be encouraged.
- **Block 3: Encourage private sector participation.** Encouraging greater private participation is an important step towards the establishment of innovative, efficient and sustainable space systems. Private actors have an important contribution to make to the development and operation of space activities. They are a key source of innovation and bring expertise and skills that are absent in the public sector. Governments need to ensure that private actors are well-placed to develop new and innovative applications.

Further details on each of the three blocks are given below.

Block 1: Implement a sustainable space infrastructure

The existence of an efficient, robust and sustainable infrastructure is essential for the smooth operation of modern economies. As eloquently illustrated when major failures occur, the world depends on the discreet, but ubiquitous, presence of such infrastructures as communication or electricity networks, as well as on their seamless and almost invisible – but essential – operation.

The same is true of space. This means that developing such an infrastructure should be a major objective of public policy.

The term "space infrastructure" is defined here as encompassing all space systems, whether public or private, that can be used to deliver space-based services. These include space segments as well as ground segments.

Two sets of recommendations are presented:

- The first focuses on that part of the space infrastructure that is "user-oriented", i.e. designed to provide services to society at large; these currently include communications, navigation and Earth observation services.
- A second set addresses efforts that governments may wish to consider to strengthen the space transport and servicing infrastructure. This is a critical enabling infrastructure that will play a major role in the coming decades in the development of the user-oriented infrastructure, and more generally, of the space sector.

The term "space transport and servicing infrastructure" is given a broad interpretation. It encompasses not only the facilities needed for launching payloads into space, and for eventually bringing them back to Earth, but also support functions, such as in-orbit servicing and the management of space debris, that will become increasingly essential for the effective operation of the user-oriented space infrastructure.

Block 2: Encourage public use

Typically, governments are major users of infrastructure, whether they use public infrastructure to deliver services to citizens or whether they use private infrastructure as an input in their activities. In most cases, public services are financed by general taxes on the population at large and are provided free of charge or at marginal cost.

Space infrastructures offer very attractive opportunities for pursuing a broad range of public missions in a cost-effective manner. In particular, space assets can help to provide solutions for long-term societal needs (e.g. environment, disaster relief, remedying the digital divide). Unfortunately, such opportunities may not be fully exploited for a variety of reasons, ranging from lack of information to technical problems, or the existence of bureaucratic rules that prevent the effective use of the infrastructures. Hence, a systematic approach to foster use is needed that fully takes into account all major impediments.

Two sets of recommendations are presented. The first focuses on actions that governments can take at national level. The main thrust is on measures designed to strengthen co-operation between user ministries and space agencies to foster the effective generation and use of space-based data and to facilitate transactions between suppliers and public users of space-based services.

The second set addresses actions that governments can take at international level to take better advantage of the ubiquity that space-based services offer. These recommendations include a broad grouping of public initiatives, ranging from risk prevention, emergency support and assistance to developing countries in the management of their resources, to the monitoring of the effective application of international treaties.

Block 3: Encourage private sector participation

While space activities were essentially public at the beginning of the space age, the role of private actors has been on the rise. They have been able to exploit successfully, in some markets, technologies that were originally developed in co-operation with or for the public sector, as in the case of telecommunication satellites. Moreover, the end of the cold war created an environment more conducive to the commercial exploitation of space. In a more open world, space firms have been able to restructure and form new alliances, and the opening of markets has benefited important segments of the industry. These commercial developments have often opened the door for more cost-effective ways to address important societal issues using space technologies (*e.g.* telecommunications networks in remote areas, Earth observation high-resolution data for disaster management).

Despite such progress, the development of commercial space remains fragile. First, the upstream segment of the industry (*i.e.* space asset manufacturing and launching services) continues to face high costs and remains very dependent on governments. Second, the development of the downstream segment (*i.e.* space applications such as satellite communications services, Earth observation services and satellite-based navigation services) is uneven. Some components remain underdeveloped (*e.g.* Earth observation) despite years of efforts, while others (*e.g.* navigation), although promising, are still at an early stage of development or under public control.

To overcome some of these weaknesses, governments need to take action so as to ensure that private actors are in an optimal position to develop innovative applications that contribute fully to the economy and society at large. Moreover, governments should take full advantage of the expertise and resources of private space actors for the development and operation of space infrastructure. In this regard, three sets of recommendations are presented:

- Recommendations for creating a more supportive legal and regulatory environment.
- Recommendations for strengthening the private provision of space goods and services.
- Recommendations for fostering a more supportive international business environment.

Introduction

Over the centuries, space has captured the imagination of many. Science fiction authors have often depicted a future in which humans would explore space, first to the Moon then onwards to Mars and other planets in our solar system and eventually to the galaxy itself. This ongoing quest for the "ultimate frontier" was expected to bring about the emergence of a new form of civilisation in which the colonisation of other worlds – and perhaps contact with extraterrestrial life – would become the main focus of human activity.

This vision of things to come has not yet materialised. Humanity has remained largely Earth-bound. While a handful of humans have been to the Moon, none has gone back since 1974; moreover, only about 400 humans have ever been in space.

One reason for this lack of progress, despite the large amounts of resources devoted to space by governments since the beginning of the space age, is the formidable technological hurdles that must be overcome. A major setback in this regard has been the inability to develop a reusable space vehicle (RLV) capable of providing relatively cheap and easy access to space. For instance, the space shuttles, which were expected to pave the way to a true RLV, have turned out to be more costly to operate commercially than expendable vehicles and not robust enough to fly often.

Although human exploration has not progressed as originally hoped, advances have nevertheless been achieved on many fronts: space telescopes and robotic exploration have considerably increased our understanding of the universe; long exposure to space has taught us how the human body responds to zero gravity over time. Progress has also been achieved in satellite communications, Earth observation and the development of space-based navigation systems.

Given this state of affairs, one may wonder what direction the development of space will take in the future. Views differ on this question.

Many still regard the exploration of space by humans as the main thrust of future space programmes. Such a vision has been articulated by President Bush as the next main step for humankind in space, first by returning to the Moon, then by using the experience acquired in this way to launch a human space mission to Mars.

Other observers of the space scene have more immediate and pressing concerns. They worry that future developments of space technology will lead to the further militarisation of space, creating a more dangerous world in which nations will largely focus their efforts on attempts to achieve "space control" (military jargon for the ability to ensure one's own access to satellite capabilities while denying space-based services to adversaries). Indeed, ever since the beginning of the space age, geopolitical considerations have played a dominant role in shaping space programmes. This is likely to continue into the future and may have far-reaching implications for the development of the space sector as a whole in coming decades.

For still others, the future of space lies largely in the development of space applications that can be used to solve problems on Earth. They note that an important trend in the evolution of the space sector over the last few decades has been the development of civil and commercial systems, as space technologies have been gradually transferred from scientific and military applications to civil and commercial ones.

Without ignoring the key exploratory and military dimensions of space activities, the main focus of this report is the future development of civil and commercial space applications. A number of questions are addressed: Which applications appear to be promising in a longer-term perspective? How is the demand for such applications likely to evolve in alternative futures? What factors may be critical for their successful development? Is the policy context in which such applications are developed supportive enough to ensure that they meet expectations? What actions need to be taken by government to ensure that the potential that space may offer is actually fulfilled?

This volume explores first in Chapter 1 the future of the space sector and of space applications with a view to identifying those that are likely to be promising in the future. This broad assessment is complemented in Chapter 2 by a more detailed analysis of the potential value of space applications for addressing a number of challenges that will confront humankind in coming decades.

While the analysis presented in these two chapters confirms the potential value of space for society at large, it remains to be determined whether this potential is likely to be realised, given the strengths and weaknesses of the space sector. Chapter 3 therefore reviews the strengths and weaknesses of space solutions from a technological and economic perspective, while Chapters 4 and 5 focus respectively on the framework conditions (institutional, legal and regulatory) that govern space activities. This provides the basis for the recommendations contained in Chapter 6.

The report also includes three annexes. Annex A presents an overview of five case studies that were conducted in the third phase of the project. These are used as an input to the analysis on framework conditions conducted in Chapters 4 and 5. Annex B lists the members of the project's Steering Group, while Annex C provides details on the experts were involved in the project.

Chapter 1

A Scenario-based Exploration of the Future of Space Applications

> *In order to assess the long-term demand for space applications, a scenario-based approach is used to explore the role that space – military, civil and commercial – might play in alternative visions of the future, with a 30-year timeframe. The scenarios take into account the interaction of three main drivers of societal change: geopolitical, economic and environmental. The results (published in full in Space 2030: Exploring the Future of Space Applications) suggest that demand is likely to be high across a range of very different alternative futures. This is partly because military space is likely to play an important role in all scenarios. But civil and commercial space applications were also found to have generally favourable prospects, primarily because they are expected to be increasingly needed to tackle many of the world's social, economic and environmental problems. Moreover, the development of non-military applications (civil and commercial) will be influenced by the evolution of military space because of the dual use nature (civil/military) of space technology.*

Introduction

The second phase of the space project focused on exploring the future evolution of the space sector with a view to identifying space applications that may be promising in the coming decades, *i.e.* applications that are likely to generate significant net social value in either the public or the private sector. This required considering how the evolution of world events, influenced by main drivers of change, may create an environment that is more or less favourable to future space activities and how this may be reflected in the demand for specific space applications.

Because of the long timeframe used in the analysis, a scenario-based approach was adopted. Indeed, when exploring inherently unpredictable futures – as is the case for the future of the space sector – the building of a range of scenarios offers a superior alternative for decision analysis, contingency planning or mere exploration of the future, since uncertainty is an essential feature of scenarios.

The overall approach involved: *i)* constructing appropriate scenarios for providing alternative visions of the future evolution of the world; *ii)* sketching out the consequences of each scenario from political, economic, social, energy, environmental and technology perspectives and drawing the implications for the future evolution of the main components of the space sector; and *iii)* assessing the implications for the future demand of specific applications.

What follows is a short presentation of the analysis conducted in the second phase of the project, focusing largely on the implications for the space sector and for the development of space applications. A more detailed treatment can be found in *Space 2030: Exploring the Future of Space Applications* (OECD, 2004a).

The construction of the scenarios

The methodology used to construct the scenarios is based on a "blueprint" widely used by futurists. It involves the following steps:

1. Define the question to be answered.
2. Identify the drivers of change with a bearing on the question at hand.
3. Analyse the trends and factors likely to affect each of the drivers of change with a view to assessing the main uncertainties that apply to their future state and define the scenario space.

4. Select in the scenario space the scenarios that will receive particular attention.

5. Flesh out the scenarios and draw their implications for the question at hand.

In applying the blueprint to the future evolution of the space sector, three main drivers of change were identified as particularly pertinent: geopolitical developments, socio-economic developments, and energy and the environment, not only because of the key role they will play in shaping future world events, but also because of their close links with the space sector.

The main trends and factors that are likely to influence the three main drivers of change over the coming decades are reviewed below, and the scenarios that were eventually selected as a result of the application of the blueprint outlined above are described.

Geopolitical trends and factors

Nation-states have played in the past and continue to play today a lead role in international relations. One key question from a geopolitical perspective is whether they will continue to do so in the future. The general view on this point is rather negative: the power they have wielded and the role they play are expected to gradually decline over time. First, secessionist movements may be in a stronger position to wrest power from central governments, and more sub-national entities may gain independence. Second, the number of failed states may rise. In contrast, international organisations, non-state actors, including multinational corporations, non-governmental organisations (NGOs) and also organised crime and terrorist groups, are likely to become more powerful.

However, nation-states are not expected to disappear. Indeed, they should remain the main focal point of international relations over the next three decades, although they will face a more complex political environment as non-state actors become increasingly active. In addition, the pecking order of nations is likely to change. Western nations are expected to lose ground overall, while new players may move ahead and become regional powers. The United States should remain in the lead, but its lead should gradually diminish and may even be challenged by China towards the end of the period. Co-operation among "lesser powers" with a common interest in the emergence of a less hegemonic world order may intensify.

On the military front, future conflicts are expected mainly to be intra-state and increasingly to involve non-state actors, such as terrorist groups or organised crime. The proliferation of weapons of mass effect (WME) will heighten concerns about domestic security in most countries. In response, the United States (which will remain the dominant military power over the period) is expected to adopt, at least initially, a dual strategy of national security (including the construction of a national missile shield) and of pre-emptive use of military force abroad. The Europeans, too, are expected to strengthen their domestic security and consolidate their collective defence. China, India and Russia may enhance their military capability as well and are likely to seek to deter intervention by the United States through their WME.

Socio-economic trends and factors

On the demographic front, world population growth is expected to slow, with all of the increase occurring in the developing world. Population ageing will become a global phenomenon, although it will be more pronounced in the West and in some non-Western countries, notably China and Russia. In developing countries, more of the population will move from rural to urban areas, creating the need for massive investment in infrastructure. More people will also migrate from the developing world to the West, creating a continuing source of tensions in Western societies as well as new opportunities. Culture should become increasingly globalised, although resistance to change and retrenchment may be strong in the more traditional societies, leading to bouts of fundamentalism. In contrast, Western societies are likely to become more secular, pragmatic and individualistic, as well as more difficult to govern, as respect for established authority declines.

From an economic perspective, the world may become a better place to live in for more people in the next 30 years. Not only is population growth slowing, but the economy may improve if the globalisation process continues, as major new technologies come on stream and spur growth. However, economic risks will tend to increase: poor governance both at national and international levels is likely to be more severely penalised by market forces, and economic shocks could have devastating ripple effects in an increasingly interdependent world. Moreover, income inequalities are expected to grow both within and between countries, while unemployment may be a major source of unrest in countries in transition. Poverty should decline overall but is likely to be increasingly concentrated in certain regions of the world, notably Africa and South Asia.

Energy and the environment: main trends and factors

The environmental outlook is bleak. As greenhouse gas (GHG) emissions may more than double over the next 30 years, a rise in the mean temperature appears inevitable, causing a noticeable rise in sea levels, more unstable weather conditions and a geographical shift of endemic and infectious diseases towards temperate zones. Implementing appropriate GHG abatement policies at international level will prove difficult because of economies' high dependence on fossil fuels and because of the externalities involved (i.e. polluters only pay a fraction of the costs they impose on society at large). Higher levels of pollution at the local level are also expected in large parts of the developing world, together with further deforestation, soil erosion and reduced biodiversity. On the other hand, OECD countries as well as a number of middle-income countries should give greater attention to environmental issues.

On the energy front, fossil fuels are expected to continue to dominate primary energy consumption. Oil will remain the leading fuel, as oil reserves should remain sufficient to meet demand. However, major investments will be required for exploration, extraction and transport. Moreover, the price of oil may become more volatile, as conventional energy reserves decrease and are more concentrated in the Persian Gulf area. Gas is likely to continue to be the fossil fuel of choice for electricity generation, because of its relatively low carbon content; hence, demand for gas is expected to rise rapidly. Demand for carbon-intensive coal should increase at a lower rate, while nuclear power will remain contentious, despite its clear advantage from the perspective of GHG emissions. Greater efforts will be made to promote the use of renewables, but their overall share of energy consumed will remain very low, as it will prove extremely difficult to steer the energy system away from fossil fuels.

An overview of the scenarios

On the basis of these trends, three main scenarios were constructed, offering three alternative visions of how the world might evolve in the future.[1] None is "more likely" than the others: indeed, all three should be considered both plausible and unlikely. They represent only three points in the set of possible futures. They illustrate the fact that the future will not be a mere extension of the present. They also offer a useful basis for reflecting on how the space sector might evolve in the coming years. A more detailed presentation of the scenarios is provided in Table 1.3.

Scenario 1 (Smooth Sailing): This is an optimistic scenario reflecting a virtuous circle involving the three main drivers. Under Smooth Sailing, the world is at peace, multilateralism and international co-operation prevail, globalisation brings prosperity to the world, notably the developing world. More specifically, this scenario points to a global world order under the benevolent guidance of international organisations in which free markets and democracy gradually become the accepted universal model for national institutions. Major contributing factors include the growth of global trade as well as the internationalisation of production worldwide. Other significant trends are progress in transport and communications and growing interest in global issues. In a favourable economic climate, international co-operation contributes effectively to solving world problems, including the alleviation of poverty. However, the environment continues to deteriorate, despite growing concerns in this regard. Moreover, various groups that feel left out or oppose the established order on ideological grounds resist what is perceived as the "westernisation" of the world. Such opposition is reflected in the persistence of terrorist actions by transnational groups, which may use "states of concern" as a strategic base for training recruits and planning actions against their enemies. Organised crime continues to be active, taking advantage of a more open world. Both groups have access to weapons of mass effect and use them to blackmail the more vulnerable governments.

Scenario 2 (Back to the Future): This is a "middle of the road" scenario. It basically describes a return to a bipolar world in which international relations are dominated by the uneasy interaction between two blocs: the United States and Europe, on the one hand, and a coalition of China and Russia, on the other. More specifically, three major economic powers dominate the world in this scenario: the United States, Europe and China. The United States remains the main power for a while but its leadership position is gradually eroded because of its relatively lacklustre economic performance. It is challenged by a rapidly growing China, which becomes increasingly confident, rejects Western values and is eager to regain, with the support of the Chinese diaspora, its historical status of "middle empire", which it considers its rightful place in the world. Russia plays an important supporting role for China, as Russian authorities also tend to resent Western criticism. Europe remains an economic giant, but it looks inward and its institutions are weak, as the extension of the EU to 25 countries has considerably slowed further integration efforts. In the face of the assertive coalition of China and Russia, Europe strengthens its ties with the United States and enhances and co-ordinates its military forces. A bipolar world gradually emerges, in which rivalry between the two blocs dominates the policy agenda in all major spheres of activities. Tensions are particularly high with respect to access to energy and other resources, as China becomes a major importer of energy and food.

Scenario 3 (Stormy Weather): This relatively pessimistic scenario describes a world in which a breakdown in multilateralism, caused by a strong divergence of views among key actors, precipitates an economic crisis that further exacerbates international relations. Economic growth is likely to be slow and concern about the environment low. In response to sharp criticism of its interventions on the international scene, the United States acts increasingly unilaterally, withdraws from any military action not justified by an effective threat to American vital interests and decides to deploy an anti-ballistic defence system to protect the US territory against limited ballistic attacks. As it largely withdraws from the international scene, ethnic conflicts multiply leading to massive migrations and terrorism. A growing number of countries acquire a nuclear capability, increasing the potential for devastating conflicts at regional level, notably in Asia and in the Middle East. Economic conditions deteriorate as the world reverts to protectionism. Growing social and ecological problems are largely ignored as international co-operation is replaced by bilateralism driven entirely by short-term *realpolitik* considerations.

Implications of the scenarios for the future of the space sector

For each scenario outlined above, the consequences for the military, civil and commercial components of the space sector are explored. For civil space, two main aspects are addressed: the impact on space exploration and science and the impact on the expansion of the civil space infrastructure. For commercial space, attention focuses on the impact of the scenario on the business environment facing space actors, the expansion of the commercial space infrastructure and the development of the space industry.[2]

Scenario 1: Smooth Sailing

In the wake of better international relations, this scenario places less emphasis on military expenditures, although the use of military space assets increases. Major progress is achieved in applying space technology to the solution of global social and environmental problems. Commercial space also expands significantly in a more open business climate.

Military space

A more peaceful world puts less priority on military expenditures. Military space budgets decline overall. However, space-faring countries outside the United States devote relatively more resources to military space as they strengthen their network-centric warfare capability.[3] Particular attention is devoted to developing a military space infrastructure in the areas of telecommunications, Earth observation (EO) and navigation for carrying out intelligence, communications, command and control (IC3) functions.

As tensions among the major space powers diminish, they increase co-operation to cope with the threat represented by states of concern and terrorist groups. In this context, major space-faring countries agree to collaborate on the development of regional missile defence systems, as required. At the same time, the United States speeds up the development of a hypersonic cruise vehicle (HCV). Europe follows suit in collaboration with Russia.

Civil space

In this scenario, all of the world's major space-faring countries co-operate actively on the development of all aspects of civil space, including space exploration and science, basic R&D for the development of space technology as well as on the expansion of space infrastructure.

Space exploration and science. An international consortium is created to develop an ambitious extra-terrestrial exploration programme, with missions to the Moon and to Mars. By 2020, a permanent international station is established on the Moon. In 2025, the first manned mission to Mars is launched. At the same time, all main space agencies co-operate actively to achieve progress in propulsion and to develop new satellite platforms.

Development of civil space infrastructure. The positive political and economic climate provides a good basis for strengthening international co-operation to deal with the world's principal societal problems. The merit of space-based solutions is increasingly recognised and the International Space Agency (ISA) is formed to facilitate such efforts.

With the help of pioneering countries such as India, the World Health Organization (WHO) actively supports the use of telemedicine in the developing world as a way to achieve its goal of "health for all in the 21st century". At the same time, an effective system for dealing with pandemics modelled on the IC3 military concept is put in place worldwide.

The WHO joins forces with the United Nations Educational, Scientific and Cultural Organization (UNESCO), the International Labour Organisation (ILO) and ISA to promote distance learning as an effective way to reduce educational inequalities and facilitate the education of the rapidly growing working age population in developing countries. Private Western firms that invest heavily in these countries participate as well. The education ministers of a core group of countries decide to create a new intergovernmental organisation, EducSat, with the aim of providing distance education services in participating countries. Membership in EducSat gradually increases as more countries recognise the merits of tele-education.

The Food and Agriculture Organization (FAO) increases its efforts to promote more efficient use of natural resources worldwide. Space assets are widely used for monitoring crops, for pest control and for precision farming. Because space-controlled precision farming has a lighter environmental footprint, the Global Positioning System (GPS) and Galileo are used to oversee the production of "green" agricultural products, with a view to satisfying more efficiently the growing demand for "organic" products.

Collective efforts to curb GHG emissions, undertaken largely under the initiative of the European Union, create a demand for space-based technologies to track emissions. The Kyoto Protocol is followed by the creation of a world environment protection agency which sets up a space-based system for monitoring the enforcement of environmental agreements in co-operation with ISA.

Commercial space

In this more peaceful world, major progress is made towards the creation of a more open environment for commercial space. The space infrastructure that supports trade and commerce is significantly upgraded. Taking advantage of liberalisation and the emergence of new business opportunities, the space industry undertakes broad restructuring at the global level. This paves the way for significantly reducing the cost of access to space and for developing new and innovative space-based services that can fully exploit the advantages that space offers over terrestrial alternatives.

Creation of a more open business environment. With regard to the institutional environment governing commercial space, progress is achieved on several fronts. First, space firms benefit from trade liberalisation and from new international rules regarding the treatment of foreign direct investment (FDI). Second, a more business-friendly regime is adopted at international level, notably for launching activities, the disposal of space objects, the protection of commercial property rights as well as the allocation of frequencies and orbital slots. Third, all space-faring nations adopt national space legislation that conforms to a uniform or model code so that definitions, liabilities and property rights can be easily determined. Fourth, the financing of space assets is put on a sound business footing.

Significant progress is also achieved regarding the liberalisation of information flows:

- Mutually agreed rules are established for regulating e-commerce and transborder data flows.
- The regulation of operators of space assets is harmonised across jurisdictions, and applications procedures are significantly simplified. Foreign ownership rules are eliminated across the board.

Restrictions regarding space technology are relaxed. Major space-faring countries agree to ease restrictions on foreign investment, export controls and technology transfers among themselves. At the same time, they put in place clear rules designed to deny access to sensitive technology by states of concern or terrorist groups.

Expansion of the space infrastructure. The favourable institutional context provides a good foundation for the extension of the space infrastructures that support the development of trade and commerce worldwide by:

- Developing a global broadband telecommunications infrastructure for which space assets play a role not only in trunking but also in "bridging the last mile", *i.e.* in connecting end users, in competition with terrestrial networks. In this more open world, the ubiquity space offers is a major selling point for space-based solutions.

- Developing a truly global positioning and navigation infrastructure for civil and commercial use that ensures full interoperability of several existing systems. The infrastructure is used for all modes of transport and greatly facilitates the expansion of civil aviation worldwide in particular.
- Developing a comprehensive global EO infrastructure that can be used for civil security as well as commercial purposes.

The development of the global space infrastructure helps to boost productivity not only directly, because of the services it provides, but also indirectly, by forcing a harmonisation of standards at the global level.

Development of the space industry. Space firms are able to restructure globally to take full advantage of economies of scale and scope, and Russian and Chinese firms play a growing role in this process. Some of these firms become the linchpins of large space consortia that operate globally and compete directly with their Western counterparts. New firms from emerging space-faring nations, notably India, Brazil and Israel, enter the industry.

Space firms engage in fierce competition. Major efforts are made to cut costs and improve the quality of services. Large R&D budgets are devoted to developing innovative space products. Some firms attempt to reduce competitive pressures through consolidation, but such initiatives are constrained by the action of antitrust authorities.

As a result of industry efforts and the expansion of commercial space, the cost of access to space is significantly reduced. In particular, the cost of manufacturing launchers is cut drastically and major advances are made in the manufacture of micro- and nano-satellites. Space tourism starts to develop after a small firm won the X-Prize in 2004 (the contest called for launching a manned craft able to carry three people to a 100 km or 62 miles altitude and return it safely to Earth twice within 14 days). Space tourism starts first on a suborbital basis and then on an orbital basis in the 2020s. Advances on the commercial side eventually converge with the progress made by the military in developing an HCV, leading towards the end of the period to the emergence of a true reusable launch vehicle (RLV).

Scenario 2: Back to the Future

For the space sector, the confrontation between China/Russia and the West which plagues international relations in this scenario leads to the emergence of three main co-operative blocs: North America-Europe-Japan, China-Russia and India-other emerging space actors. Closer links between North America and Europe result in an integrated space industry. Space firms benefit from higher military space budgets but suffer somewhat from a less open trade and investment climate. Civil space largely devotes its efforts to reducing the cost of the welfare system through the development of dual-use technologies, while prestige activities are designed to strengthen "soft power".

Military space

Growing tensions between the West and China/Russia lead eventually to a new type of space race and the gradual "weaponisation" of space. This involves the deployment of national ground-based missile defence systems, including advanced surveillance and warning systems, first by the United States and then by other major space-faring countries. The increasing weaponisation of space is also reflected in the development of anti-satellite (ASAT) systems, including airborne and ground-based lasers and parasitic satellites, and finally the deployment of space-based lasers capable of attacking both missiles and satellites towards the end of the period.

EU countries strengthen their common security and defence policy. Military space plays a central role and a core group of like-minded countries agree to co-ordinate their military space programmes so as to minimise duplication. This leads to the rationalisation and development of Europe's military space infrastructure. The Europeans want to establish an independent space capability, but they also stress interoperability with US military space-based assets. The military space industry of the United States and the EU becomes increasingly integrated.

China gives high priority to the modernisation of its armed forces, with the support of technology transfers from Russia. It views the use of space as being of central importance and paving the way for its own network-centric warfare concept. Military build-up by the two main blocs encourages other countries, in particular India, to enhance their military space capability. The demand for communication and EO satellites increases.

Civil space

Because of international rivalries, a large share of civil space budgets is devoted to projects likely to create "soft power" in the form of additional prestige at home and abroad or as a way to strengthen or extend international influence. This environment is particularly favourable to new exploration programmes, technological developments and space-based responses to regional social demands (*e.g.* telemedicine).

Space exploration and science. Countries step up their respective exploration programmes for reasons of prestige.

The United States, Europe and Japan launch an ambitious unmanned Mars exploration programme, while phasing out the International Space Station (ISS).

The objective is to put humans on the red planet by the mid-21st century. Following some European efforts in the early 2000s, Japan initiates a lunar project, starting with the Lunar A and Selene projects, to survey the Moon's resources and prepare further exploration of the Moon.

China also initiates an ambitious Moon project, starting with an unmanned lunar probe. Russia's unparalleled experience with long-duration human spaceflight gives China an edge over the West. The official long-term goal of both China and Russia is to exploit the Moon's potential mineral and energy resources for the benefit of humanity, and the short-term goal is to increase their national prestige, both at home and abroad. By the end of the 2030s, China and Russia establish a manned outpost on the Moon.

India, with an unmanned mission to the Moon by 2008, is not far behind. The mission's aim is to showcase the country's scientific capabilities, to excite the younger generation and to increase national confidence. Going to the Moon is also perceived as an important step strategically and economically. Following the Indian model of space development, many countries place special emphasis on projects using small satellites and available technology to perform specific economically useful missions.

Development of civil space infrastructure. In this period of high social demand, space applications increase and provide government-sponsored solutions. New dual-use technologies are developed.

In the face of escalating health-care costs, telehealth, including space-based applications, offers an attractive way to deliver health services, notably homecare services to the elderly, by taking advantage of the direct-to-home (DTH) broadband capability already in place in many homes. In this way, the health status of the elderly can be continuously monitored. Tele-consultations can be made with the assistance of nurses specifically trained for this purpose. Substantial savings are realised as costly and time-consuming home and hospital visits are cut to a minimum. Other civil space efforts focus on the environment. Although little progress is made regarding the abatement of GHG emissions, the two main geopolitical blocs vie for the allegiance of developing countries by contributing to efforts to monitor pollution via satellite and providing emergency services in case of major natural disasters. Moreover, space assets are used to verify the application of regional pollution abatement treaties.

Recommendations to avoid more orbital space debris are regularly presented to existing multilateral bodies by developing countries, but little progress is achieved because the guidelines enacted are not enforceable.

Concerning technology development, significant advances in artificial intelligence (AI), robotics and nanotechnology contribute to cut the cost of space missions, as in Scenario 1.

Private actors develop suborbital launchers, and governments fund space-plane technologies as a priority for military purposes.

The end of the 2030s sees the development of energy relay satellites developed through co-operative regional efforts involving energy companies and governments. Some security concerns, in particular the vulnerability of such systems to ASAT weapons, limit their use.

Commercial space

In a tense international situation, where regional blocs tend to pursue their own strategic interests, commercial space activities tend to develop more slowly than in the first scenario. A limited but real return to protectionism in the space sector is encouraged by security concerns. Each bloc develops commercial applications to meet its own strategy.

The business environment. In this scenario, internal space markets are largely protected, and many advances mentioned in Scenario 1 do not take place. Technology transfers between blocs face high regulatory hurdles.

The financing of space activities becomes easier as the UNIDROIT protocol on space assets is finally approved, but only in western countries (see Chapter 5 for details concerning UNIDROIT).

A North Atlantic Free Trade Agreement is established between North American Free Trade Agreement (NAFTA) countries and the European Union. A number of Latin American and North African countries are associated to the agreement. This helps to create a more open trading environment across the Atlantic and fosters the development of space applications to some extent. However, the emphasis on military space tends to slow the development of commercial space, as space firms devote a higher proportion of their resources to military contracts. Moreover, the military are reluctant to transfer technology to the private sector for fear that it might fall in the wrong hands.

In China, the two-track space strategy adopted in the first decade of the 21st century (establishing joint ventures with Western firms and participating in major co-operative efforts) allows it to expand its technological expertise and know-how and to gain commercial space independence (i.e. the ability to offer space products and services without first obtaining agreement from Western providers of key components) before breaking with the West. China's increasing co-operation with Russia and some Association of Southeast Asian Nations (ASEAN) countries permits it to export space-based services (e.g. space imagery, access to telecommunications satellites) to the rest of the world. As a large country with limited infrastructure, it finds satellite transmission a useful way to develop a nationwide telecommunications network quickly. China is also keen to use space imagery to find and manage natural resources and to export its competitively priced space products to the rest of the world. Because of growing international tensions, China is forced to target its exports to countries outside the Western free trade region and to compete with other important emerging players (e.g. India), as Western firms are subjected to stricter export controls on space assets and components considered strategic by military experts

Limited expansion of the commercial space infrastructure. Many new space-related products and services are developed regionally. However, export and investment restrictions tend to reduce the broad diffusion of new technologies and applications.

Restrictions on information flows (e.g. Internet regulations, operator licensing) negatively affect the telecommunications sector and the development of "infocom" applications in some countries. The broadcast industry (e.g. television via satellite) faces strong regional competition from cable operators.

The use of space-based navigation systems is widespread for all forms of transport, and notably for civil aviation. This situation forces developers of navigation systems (United States, Europe, China, Russia) to co-ordinate their efforts and to discuss interoperability issues.

The growing demand for energy results in further exploration (e.g. oil, gas) and greater need for appropriate space-based technologies. In this regard, the improved hyper-spectral capability of remote sensing proves to be particularly useful for oil exploration. Space assets are also used extensively to monitor pipelines and to assist in major energy infrastructure projects, which are needed to meet the rising demand for energy. However, rivalry between the blocs results in substantial duplication of effort.

A new commercial sector, suborbital space tourism, sees some limited development, especially in the West. Enthusiastic private entrepreneurs develop suborbital launchers with off-the-shelf dual-use technologies, but the tense international security environment restricts their commercial activities (e.g. launch conditions, number of flights per year). Based on the commercial success of new adventure tourism activities, some companies seek to use the new military space-plane technologies.

The development of the space industry. In the United States, Europe and Japan, semi-private space firms further integrate their activities and take advantage of higher military budgets to develop dual-use applications under public-private partnerships. In their own markets, they are able to gain some protection against cheap imports from other players (e.g. China, Russia, India).

In each block, space companies compete with each other, but they also all face strong competition from regional terrestrial systems.

Scenario 3: Stormy Weather

Under this scenario, multilateralism breaks down and security, defence and other strategic government uses of space become increasingly important. Most space powers tend to develop their systems independently, forming alliances as needed, but this is a divided world with no clear alliances. Civil space tends towards dual-use technological developments or activities that contribute to enhance soft power. However, the value of the space infrastructure for the effective and efficient delivery of social services is increasingly recognised in major space-faring countries. The impact on space business is mixed. On the one hand, space firms benefit from government contracts and the spin-off opportunities they offer. On the other hand, markets become more fragmented, export controls more stringent and the restructuring of firms at international level is stifled by national security considerations.

Military space

In a world perceived as increasingly hostile to the vital national interests of important space powers, the military space budget increases worldwide.

In the United States, although specific programmes are modified by successive administrations, efforts to militarise space win out. The United States decides in the early 2000s to carry on with its anti-ballistic programme, as the abandonment of the ABM treaty in June 2002 effectively removed restrictions on placing weapons in space.

The United States steps up its efforts to develop an unmanned reusable hypersonic cruise vehicle capable of reaching speeds of Mach 10-15 and of placing ordnances on military targets anywhere in the world within a few hours. The HCV becomes fully operational in the late 2020s for military purposes. In addition to striking targets around the globe, it can also be used for launching short-term satellites to bolster communications, or for remote sensing or navigation in a target region. Smaller versions of the HCV are developed.

Following the US lead, a growing number of countries decide to develop or strengthen their own military space assets, including for communication, Earth observation and navigation.

After lengthy discussions, Europeans finally launch a major military space programme by the end of 2010s. The programme is designed to reduce the large and growing gap in military space capability with the United States and to keep up with the efforts of other major space powers, notably China. Europe develops this full-fledged military "system of systems" for space activities to ensure its independence and its autonomous and informed decision-making.

China also gives high priority to strengthening its military space capability over the period, owing to its volatile partnership with Russia. In response to US missile defence deployment, China develops and ground-tests an advanced anti-satellite weapon. It also strengthens its effort to develop a ground-based laser.

Russia tends to use its industrial and scientific know-how to partner with the most promising ally in military space activities. Its numerous military exports allow it to develop, in a limited way, new technologies, especially in propulsion, which are of great interest to one-time potential partners (*e.g.* China, Europe).

Civil space

Because of the depressed economic conditions, there are strong pressures on discretionary budgets, notably on programmes that are not perceived as being of immediate benefit.

Space exploration and science. No major common international exploration programmes are pursued, as national and regional programmes remain in the forefront. Space agencies undertake strategic co-operative efforts, essentially to take advantage of and to influence the research efforts of other nations. However, some efforts backfire and/or conflict with strategic objectives. For instance, the ISS programme stalls by the late 2010s, as concerns in the US Congress about technology transfers to Russia result in budget cuts for this activity. This leads eventually to a winding down of ISS activities in a climate of mistrust.

Some countries try to strengthen their soft power through a number of spectacular initiatives designed to demonstrate their space prowess to the world. As in the second scenario, these efforts largely take the form of competitive missions to the Moon and to Mars. However, the scientific value of these space ventures is undermined by duplication of effort and by the priority given to technology over science. Moreover, the missions are less spectacular in this scenario because significantly fewer resources are available.

The expansion of the civil space infrastructure. Even though civil budgets are quite limited, some countries still recognise that civil space programmes are not just a cost item but can sometimes be considered an investment that contributes significantly to their development. National civil space research efforts are largely devoted to the development of dual-use technology. This applies notably to meteorology, Earth observation, telecommunications and navigation systems as well as to launchers.

China and India, large countries with limited terrestrial infrastructure, lead the world in the development of space-based telemedicine and distance education applications and are able to export their expertise to other developing countries in Asia, Latin America and Africa. The Indian model of autonomous space development inspires many emerging space powers (*e.g.* Brazil, Turkey).

In Europe and North America in particular, the development of DTH offers a convenient and cost-effective platform for applications designed to reduce the digital divide and to promote home health-care services outside major urban areas in order to alleviate the growing pressure on health-care budgets.

Commercial space

Government attention to military space has some positive effects on commercial space. However, those effects are offset, in part, by the deleterious consequences of poor economic conditions and market fragmentation.

The business environment. As in Scenario 2, protectionism tends to be quite strong, limiting technology transfers and export possibilities. Some selected lucrative export markets for space products and services remain open, as a growing number of countries are keen to build a space capability and to acquire the necessary technology from major space powers. Such powers agree to do so for selected countries for strategic reasons and to extend their regional influence.

Private investment in space is cut back, as high-risk investment opportunities requiring the raising of large up-front capital are the first to be postponed when economic conditions are depressed. The poor investment situation is partly offset by the decision of a number of governments to purchase space services directly from private sources rather than to create them within government agencies. However, of necessity, these profit opportunities are heavily regulated and dependent on the budget process.

Limited expansion of the commercial space infrastructure. On the positive side, advances for military space provide spin-off possibilities for civil and commercial space applications. However, little progress is made in developing other segments of commercial space. In particular, strong regional barriers to information have very damaging impacts on telecommunications services (*e.g.* television via satellite, Internet).

Some space assets (*e.g.* remote sensing, navigation systems) are used extensively for monitoring the production and distribution of oil and gas (navigation and Earth observation systems). Substantial exploration also takes place in some countries anxious to reduce their dependency on imports.

In the launch sector, the early development of a civil and commercial version of the small launch vehicle (SLV) and of the HCV gives the United States a strong comparative advantage for launching small satellites towards the end of the 2020s. However, protectionist measures in other countries prevent the

US industry from fully exploiting its technological advance. This does not prevent a number of less developed countries (LDCs) from taking advantage of the cheaper launching fees offered by US firms to send their satellites into space.

Suborbital space tourism develops more slowly than in Scenario 2, amid strong international tensions. The general environment of distrust and the dual nature of launchers strongly limit space tourism's commercial possibilities.

Development of the space industry. As in Scenario 2, most space companies face strong internal competition in their respective regions. The relative progress in space technologies, due to the high priority accorded to military space, gives space operators an edge over their terrestrial competitors in some cases (*e.g.* surveillance systems). This helps commercial providers of space-based services to maintain revenues in a depressed market. However, space systems in direct competition with terrestrial alternatives (*e.g.* cable operators) suffer major losses of revenues, as markets become increasingly fragmented.

Conclusion

The three synthesis scenarios presented here provide very different future visions of the world, ranging from the optimistic outlook of Smooth Sailing, which foresees major advances to improve human conditions in a spirit of international co-operation, to the dark picture depicted by Stormy Weather, which sees a world caught in a vicious circle of violence and in which most of the major problems facing humanity today (*e.g.* conflicts, poverty, malnutrition, disease, environmental degradation) become worse. Even the more optimistic scenario is not without its darker side, notably the rise of non-state actors increasingly capable of using WME in the pursuit of their cause, whatever it may be. Despite these differences, the scenarios share some common ground with respect to their impact on space.

Military space plays an important role in all three scenarios, although in different degrees. Even in the relatively peaceful world of Smooth Sailing, security concerns are high and a number of countries are anxious to strengthen their military space capability. This results in a strong and robust demand for military and dual-use space assets worldwide, as well as substantial increases in military and dual-use R&D budgets for space outside the United States.

Civil space also plays an important role in all scenarios, although for different reasons. In Smooth Sailing, its role in fostering international co-operation to solve world problems (education, health, environment) is central. In Back to the Future, prestige projects and attempts to increase soft power give importance to spectacular ventures to the Moon or to Mars. Space is also called upon to solve world problems but in a less co-ordinated, more fragmented and less effective manner. Even in Stormy Weather, the outlook for civil space is not bleak, although the resources devoted to it may be quite small. As in the other scenarios, the development of dual-use technologies remains a priority; prestige and soft power are also important drivers. World problems are addressed in a more fragmented manner than in Back to the Future, but important gains can still be made if space firms are able to demonstrate that space solutions can bring about major savings for cash-strapped governments.

Commercial space varies much more than military space across the scenarios. It thrives in the Smooth Sailing scenario, remains strong in the Back to the Future scenario but is seriously constrained in the Stormy Weather scenario. It is worth noting that for space firms in Europe and the United States, Scenario 2 may be the most favourable because of the protection it offers against competition from non-Western firms. In all three scenarios, commercial space benefits from military budgets for space.

The following section discusses how the importance of space in the various scenarios may translate into a demand for space applications and – leaving supply-side considerations aside – draws some conclusions about which applications may be viewed as "promising" from a demand perspective.

Implications for space applications

The demand for space applications considered here takes into account private or "commercial" demand, "social" demand as well as military demand. The analysis is essentially qualitative; no attempt is made to quantify demand.

When attempting to identify "promising" applications, their chronological development as well as the interrelations of applications along the "space value chain" must be considered. The space value chain is made up of three broad groupings of activities or services: information services, transport services and manufacturing. In terms of the chronology, "weightless" applications such as information applications are likely to be developed first, given the high cost of access to space. Transport applications would follow since they rely heavily on information applications, notably for communication and navigation. Manufacturing/mining applications, which depend on the effective development of the first two groupings, would be expected to come last. The question then is: How fast will this sequence occur and how will each of the groupings evolve over time? This is roughly summarised in Table 1.1. The cost of access to space is purely indicative; it is intended to give an idea of when a particular group of applications is likely to become commercially feasible. For instance, space tourism might start to become viable when the cost of access to space declines to USD 1 000/kg, assuming that the reliability of space flights increases by several orders of magnitude.

Table 1.1. **Broad categories of space applications**

	Sub-categories	Cost of access to space
Information services	• Communications • Earth observation • Navigation	~ USD 10 000/kg
Transport	• Public access to space • Space transport	~ USD 1 000/kg
Manufacturing	• Solar energy • Microgravity • Lunar extraction	~ USD 100/kg

The following sections consider how the implications of the scenarios for the three main components of the space sector affect potential demand for various space applications.

Potential future demand for information applications
Telecommunication services

All the scenarios presented above suggest that, overall, the potential demand for telecommunications services should remain strong in a broad range of possible futures. The development of satellite broadband in the coming years (notably fourth-generation broadband) appears inevitable, although it occurs at different rates across scenarios, may involve different actors and may only represent a niche market. Three main factors are at play. First, the overall rate of economic growth drives commercial and social demand for telecommunications services. Second, the degree of fragmentation of markets affects the relative competitive position of space-based solutions and their terrestrial alternatives. Third, the level of international tensions drives military demand for space-based telecommunications, notably broadband, in the context of the development of network-centric warfare capability.

It follows that from a social and commercial perspective, space-based solutions are strongest in Scenario 1 and weakest in Scenario 3. In Scenarios 1 and 2, space operators may be able to leverage their strong position in direct broadcast satellites (DBS) to extend their services to broadband users. Their ability to do so will depend however on whether the broadband solutions they will offer are attractive enough, when compared to terrestrial alternatives (*e.g.* cable, ADSL). They are also well placed to provide the "last mile" solution for rural and remote areas and address the needs of individuals on the move. In Scenarios 2 and 3, the relative decline in market demand is offset, at least in part, by an increase in military demand.

In all scenarios, the demand for mobile communications should be high. Space can capitalise on this if appropriate technical solutions are found. However, space telecommunications face serious terrestrial competitors and could largely lose out to fibre optics (or future terrestrial alternatives) in urban areas or for communications between urban areas.

The trends towards a more mobile society and the increasing costs of transport are strong drivers in favour of distance learning and telemedicine, in both the OECD and the non-OECD areas. Their development could help to reduce the "medical divide" and the "digital divide" within and among countries. Even if economic growth slows (Scenarios 2 and 3), these applications may remain attractive owing to their cost-saving features (*e.g.* extension of homecare services). Finally, large multinational enterprises (MNEs) are likely to take advantage of the opportunities that distance learning offers to train their staff and keep their skills up to date. The military will be interested in both fields of application. In this context, space-based solutions may play an important role, not only in rural and remote areas but also in urban areas, depending on how technology evolves. The increasing mobility of the population should also favour space-based solutions.

Earth observation services

Earth observation is an aspect of space applications that is technologically mature and very valuable from the military, social and commercial perspectives. Many applications are being developed, building on specific tools and techniques, such as remote sensing imagery, geographic information systems (GIS), digital terrain mapping (DTM) and subsidence monitoring. Although alternative technologies (*e.g.* aerial observation) have progressed and new ones are in the wings (*e.g.* unmanned vehicles), space-based observation has a unique capability to provide the "big picture" and is becoming increasingly flexible. Significant progress has also been made in the systems needed to exploit the data collected by EO satellites.

From a military perspective, EO is a critical component of IC3, notably the intelligence and control elements. For instance, it offers a unique capability to monitor the deployment of hostile forces or to provide in real time a picture of the progress achieved on a particular theatre of operations. EO has also proved to be an effective tool for monitoring the application of disarmament treaties.

From a civil perspective, EO has a wide range of applications in support of important public responsibilities, including security (natural disaster prevention and management, search and rescue missions), the management of natural resources, land cover and urban planning, weather forecasting and climate change monitoring (*e.g.* as addressed by the Global Monitoring for Environment and Security [GMES] programme).

From a commercial perspective, EO could be used in the future by a growing range of businesses from insurance companies wishing to estimate the cost of a natural disaster to farmers who want to know the potential size of a particular crop or wish to apply precision farming techniques.

Demand for Earth observation is expected to increase in all scenarios, although the composition of demand varies. For instance, military demand will likely be stronger in Scenarios 2 and 3 than in Scenario 1, while civil and commercial demand is likely to be stronger under Scenario 1. Finally, applications related to strengthening domestic security (including dealing with natural and man-made disasters and extreme weather conditions) should be high under all scenarios. The main difference across the scenarios is the degree of international co-operation in the development of systems. They can be expected to be more international and complete in terms of coverage, hence more effective, in Scenario 1 and to be more fragmented, involve more duplication and be less effective in Scenarios 2 and 3.

The important public-good and commercial elements of Earth observation provide strong incentives to develop user-oriented applications over the next 30 years. Aside from the obvious military applications (*e.g.* surveillance), space observation can provide solutions to a number of key social and industrial problems:

- Demand for energy is expected to increase worldwide. A greater need for appropriate space-based exploration applications (*e.g.* for oil, gas) is probable, as remote sensing's increasingly improved hyper-spectral capability is well adapted to oil exploration.
- National, regional and/or international security-related programmes (*e.g.* weather, environment, disaster prevention systems) may be developed. There is significant potential demand for space-related applications for natural disaster prevention and management.
- Some treaty monitoring activities from space for environment and/or disarmament as well as for the verification of the application of national or regional policies (*e.g.* Common Agricultural Policy) could be put in place. In certain cases, space verification of agreements with billions of dollars at stake will be the only or the main tool for this purpose (*e.g.* enforcement of GHG emissions abatement accords).
- Monitoring and management of land cover for urban planning, forestry management and agriculture will be an increasingly important task for local and regional decision makers, as well as commercial actors, who need to improve safety, profitability and the environment (*e.g.* formulation of land-use planning proposals, monitoring urbanisation, insurance assessments, precision farming).

Some of the relatively high potential demand for such applications could be met by terrestrial competition, as non-space systems (*e.g.* aerial photography) may also benefit from advances in electronics and other sectors and provide alternatives for some uses.

Positioning and navigation

Satellite radio navigation is based on the emission of signals from satellites that give an extremely precise indication of the time. With the use of a small cheap individual receiver, one can determine one's own position or the location of any moving or stationary object (*e.g.* a vehicle, a ship, a herd of cattle, etc.).

Originally developed for military use, space-based positioning and navigation services have found a growing range of civil applications in recent years. These include assistance to the movement of people and goods in various forms of transport (road, rail, aviation, public transport, maritime), civil protection, management of natural resources (*e.g.* fisheries), development of land infrastructure (*e.g.* energy networks), urban planning and keeping track of moving objects.

The demand for positioning and navigation services is expected to be strong under all three scenarios, although the composition of demand may vary somewhat. For instance, military demand is likely to be strongest in Scenarios 2 and 3, while commercial demand is higher in Scenario 1. More rapid development of urban infrastructure and land transport networks in Scenario 1 should create a strong derived demand for space-based positioning services by the construction industry and urban planners. Expected increases in traffic should also generate substantial growth in the demand for navigation and location-based services. The main difference across the scenarios relates to the infrastructure that is eventually put in place, which may be fully interoperable in Scenario 1 but very partially or not at all interoperable in Scenario 3, with considerable differences in the quality of the positioning and navigation services offered and hence their value for users, notably in urban areas.

The increase in the mobility of individuals and goods will be particularly significant and will require a major upgrade of transport infrastructure, notably for air transport, road transport and public transport. Positioning and navigation systems will play a key role for the development of the necessary infrastructure, the management of the growing volume of traffic and the operation of aircraft and vehicles. In particular, the international air traffic management system should rely heavily on space-based navigation systems in the next 30 years.

At the same time, the integration of positioning receivers with mobile phones will provide opportunities to create a multitude of consumer-oriented location-based services (LBS) that offer positioning, direction finding, real-time traffic information, etc.

The potential market for LBS applications is enormous, as it is linked to the expansion of the mobile phone market. Market forecasts indicate that 2.7 billion mobile phones will be in use worldwide in 2020.

Potential future demand for space transport and manufacturing

Space tourism/adventure

Space tourism, or rather space adventure, is an application that involves taking paying customers to space, on either a suborbital or orbital flight. Suborbital flights involve a short excursion above 100 km. Full-blown space tourism/adventure implies the organisation of longer trips to space, possibly with a limited stay in orbital facilities. It is expected to be attractive to people who are willing and able – despite the expense – to go to extreme lengths to live extraordinary adventures.

According to the World Travel and Tourism Council (WTTC), an international body representing the private sector in all parts of the travel and tourism industry, tourism is one of the world's largest and fastest-growing industries, representing more than 10% of the world's gross domestic product (GDP). In 2003, the tourism sector was expected to generate USD 4 544.2 billion. Over the next ten years, it could grow by 4.6% a year in real terms, i.e. to an estimated USD 8 939.7 billion in 2013.

Adventure tourism is an increasingly profitable segment. Trekking in isolated lands, safaris and mountain climbing are being complemented by rides in military jets. For instance, treks to the top of Mount Everest are increasingly popular despite the dangers involved, the costs (licence costs alone reach USD 50 000), and a six-year wait. Space tourism may become the next step in adventure tourism, even if the possibilities are limited at first. This extension of tourism and travel to space is present in all three scenarios, but the demand foreseen differs, as it depends significantly on international tensions, security imperatives and development of space transport.

Various studies of the potential demand for space tourism have been carried out over the years. One study conducted in 2001 for the National Aeronautics and Space Administration (NASA) under the Space Launch Initiative (SLI) concluded that at a price of USD 400 000 a ticket, 10 000 passengers a year would purchase a trip to space, generating an annual USD 4 billion. An SLI-funded study carried out in 2002 suggests that by 2021, the orbital segment might involve 60 passengers a year and yield revenues exceeding USD 300 million. In addition, the suborbital segment might attract as many as 15 000 passengers a year, for revenues in excess of USD 700 million.

The logical first step described in all three scenarios is the coming of age of suborbital tourism as an adventure tourism activity. The customer base tends to decrease across scenarios as the general geopolitical and economic environment deteriorates. Full-blown space tourism in orbit is only considered in Scenario 1, as economic conditions improve and civil/commercial use of military technologies accelerates. In this scenario, the advances realised on the commercial side eventually converge with the progress made by the military in developing reusable technologies, leading towards the end of the period to the emergence of a true RLV. In Scenarios 2 and 3, the general environment of distrust and the dual nature of launchers strongly limit space tourism's commercial possibilities.

In all three scenarios, there is a strong drive to reduce the cost of access to space, notably by developing a genuine RLV. In Scenario 1 both the military and business entrepreneurs have an incentive to develop such a vehicle. In Scenarios 2 and 3, the main driver, at least initially, is the military. As countries co-operate more for civil goals, the development of space tourism is facilitated.

Space production activities

In the context of this study, space production includes three types of activities: in-orbit manufacturing (*e.g.* testing and manufacturing of pharmaceutical products and new alloys in microgravity), space power generation (*e.g.* development of space solar power systems to provide energy from space to Earth) and extraterrestrial mining (*e.g.* mines on the Moon).

There have been some in-orbit manufacturing activities over the past decades. They mainly consist of scientific and limited commercial research concerning pharmaceutical products and materials on different space platforms. The demand for larger-scale space manufacturing in microgravity remains largely potential and hypothetical. It may eventually emerge for very high-value items (*e.g.* crystals for semiconductors, new alloys and composites) if the cost of access to space is significantly reduced.

The demand for space power generation is also quite hypothetical, although it is addressed in all three scenarios. Current terrestrial energy supplies should remain sufficient to meet demand over the next three decades. However, there is growing social demand for cleaner energy sources. It is possible to envisage space-generated power systems complementing classical energy sources in time. Theoretically, the economic potential exists, but the ability to produce energy in space and transmit it to users on Earth at a competitive price is far from technically feasible at present. On the other hand, there may be opportunities to use space power satellites to meet the demand for energy consumption in space. Moreover, relay satellites for transporting energy from producers on Earth to consumers on Earth might become feasible over the period.

Mining extraterrestrial bodies (*e.g.* the Moon, asteroids) to provide new resources for Earth or build on-site outposts, is an activity for which potential demand is not well defined. However, in the next 30 years, it may evolve from scientific and exploration missions. There may be future commercial opportunities, but technical and regulatory hurdles are important.

The overall development of the space production sector will depend critically on a drastic reduction of the cost of access to space, on the availability of cheap and reliable sources of energy in space as well as on the evolution of space production processes and techniques. It will also depend on the advantage that producing in space may offer over producing on Earth. So far, such an advantage has not been demonstrated.

In-orbit servicing

In-orbit servicing includes servicing of space platforms (*e.g.* satellite, space station) for replenishment of consumables and degradables (*e.g.* propellants, batteries, solar array); replacement of failed functionality (*e.g.* payload and bus electronics, mechanical components); and/or enhancement of the mission (*e.g.* software and hardware upgrades). It should also logically include the orderly disposal of satellites at the end of their useful lives, as well as the management of space debris.

Up to now in-orbit servicing has been limited to manned missions (*e.g.* shuttle mission to repair the Hubble telescope) and software upgrades (*e.g.* Galileo mission). The main limitation is cost and the fact that satellites are typically not designed with servicing in mind.

The potential demand for in-orbit servicing and for the disposal of space debris is likely to increase under all scenarios. The ability to service satellites would enable operators to provide more reliable service with less need for expensive back-up satellites and allow them to keep their spacecraft's electronics up to date. The weight and cost of satellites could also be reduced if refuelling is easier and cheaper. The military are likely to be a major driving force behind the development of in-orbit servicing as a way to keep their fleet of expensive satellites in low orbit fully operational. In-orbit servicing could also help to reduce the threat of anti-satellite systems. Moreover, as the orbits that are "homes" to many of today's and tomorrow's space systems are increasingly crowded with space debris, there may be strong strategic, social and commercial demand for the orderly disposal of satellites.

Individual or joint space exploration programmes to the Moon and Mars by different space powers (United States, China, Russia, Japan, Europe) may stimulate the development of some space infrastructure and encourage in-orbit servicing practices. The main limitation relates to the development of service satellites (*e.g.* NextSat) and of an appropriate infrastructure to carry out the servicing.

Conclusions regarding demand

An analysis of the scenarios presented here reveals some general demand trends for each major sector of space applications: telecommunications, Earth observation, navigation and potential new sectors (space production, space tourism). Potential demand reflects social, governmental and commercial imperatives but may be altered by various factors (*e.g.* competition from terrestrial applications). In light of the above discussion, Table 1.2 shows a number of applications that seem to have relatively strong potential demand over the next decades.

Table 1.2. **Potentially promising applications**

- Distance education; telemedicine.
- E-commerce.
- Entertainment.
- Location-based consumer services.
- Location-based services: traffic management.
- Land use: precision farming and natural resource management (forest, water, energy, etc).
- Land use: urban planning.
- Land use: exploration (*e.g.* oil).
- Disaster prevention and management.
- Environment applications and meteorology.
- Monitoring of the application of treaties, standards and policies.
- Space tourism/ adventure (suborbital and orbital).
- In-orbit servicing.
- Power relay satellites.

1. A SCENARIO-BASED EXPLORATION OF THE FUTURE OF SPACE APPLICATIONS

Table 1.3. **The synthesis scenarios**

	1. Smooth Sailing	2. Back to the Future	3. Stormy Weather
Political	Co-operation increases at international level as countries realise that, in an increasingly inter-related world, independent action is more and more constrained. However, the threat of the use of WME by terrorists and criminal groups remains high.	China takes an increasingly confrontational posture with the United States and the West. Russia, creates stronger links with China. The two complementary economies become increasingly integrated. Western countries remain united. Both blocs respond to growing tensions by strengthening their military capability.	Confronted with terrorism and other threats to its national security, the United States becomes increasingly isolationist. This undermines the authority of the UN, encouraging a growing number of larger countries to follow suit.
Economic	Rapid progress in a broad range of technologies fosters high rates of growth worldwide, particularly in developing countries which gradually catch up with the West. WTO discipline is strengthened while intellectual property and FDI are better protected.	Sluggish economic growth prevails in the West. In contrast, China enjoys high sustained rates of growth. China's rising prosperity results in a large increase in its demand for food and natural resources, including oil. China's efforts to capture a larger share of supplies outside Russia result in confrontations with the rest of the world, notably the West.	The breakdown in the multilateral regime is reflected in a gradual erosion of WTO discipline. When confronted with economic difficulties, countries do not hesitate to adopt "beggar thy neighbour" policies, thereby provoking retaliatory actions by their trading partners.
Social	Growing prosperity provides the resources to deal with the costs associated with an ageing population in developed countries, while in developing countries it generates job opportunities for the rapidly growing labour force. International co-operation and economic prosperity also provide the basis for dealing more effectively with poverty and malnutrition.	Slow economic growth in the West exacerbates social tensions. Immigration is viewed with greater hostility, and more emphasis is placed on law and order in ageing societies. Social tensions in the South are alleviated by the economic boost caused by the diversion of western trade and investment from China to them.	Political tensions and economic difficulties are reflected in serious social tensions in both the West and the rest of the world. Security concerns move to the top of the policy agenda. Poverty is on the rise in the South, and migratory flows to industrialised countries increase significantly, further exacerbating these countries' social and political problems.
Energy	Rapid economic growth results in an increase in the demand for energy. However, tensions over energy remain low as alternative sources of energy (e.g. tar sands, renewables) are developed and as market and other mechanisms promote more efficient use of energy.	Heavy dependence on fossil fuels continues. The main priority remains high growth, low energy prices and stable supply, in both the West and the rest of the world. Concerns about security of supply rise and major efforts are made to develop alternative sources of energy.	Slower growth results in reduced increase in energy demand. However, security of supply is of primary concern for most countries, thus exacerbating tensions among energy-importing countries. This also helps to spur efforts to find alternative sources of energy.
Environmental	Environmental problems increase in the medium term. But, as more countries reach medium-income status, more countries clean up local pollution accept limits on their GHG emissions.	No international agreement on the control of GHG emissions is reached. The environment deteriorates. However, co-operation to deal with local pollution problems increases at regional level.	Protection of the environment is not a priority, given high public concern about national security, economic development and security of the energy supply. Pollution is reduced in OECD countries, as well as in some medium-income countries.
Technology	Progress in information technologies, biotechnology and nanotechnology spurs economic growth and provides new ways to deal with environmental problems.	Innovation in the West is slowed by poor economic conditions. Technology transfers to the South favour Western-friendly nations. Priority is given to military research.	Innovation is slow except in the field of military technology. Diffusion of new technologies to developing countries is limited.

Notes

1. See OECD (2004) for more details on the methodology used and the experts consulted to build the scenarios.

2. In order to make the scenarios as realistic as possible, it is necessary to refer to specific concepts and space developments, which are treated in more detail in subsequent chapters (*e.g.* developments in space transport systems, trends in the development of specific applications and their legal regime). Cross-references are included for specific items in order to clarify these concepts for non-space experts.

3. The term "network-centric warfare capability" refers to the configuration of armed forces in which all units as well as individual soldiers are interconnected by a multi-layered communications network which enables commanders to monitor action in the battlefield and give orders in real time.

Bibliography

Antón, P.S., R. Silberglitt and J. Schneider (2001), "The Global Technology Revolution: Bio/Nano/Materials Trends and Their Synergies with Information Technology by 2015", Rand Corporation, Santa Monica, California.

Baer, W., S. Hassel and B. Vollaard (2002), "Electricity Requirements for a Digital Society", Rand Corporation, *www.rand.org/publications/MR/MR1617/*.

Baker, J., K. O'Connell and R. Williamson (eds.) (2001), Commercial Observation Satellites: At the Leading Edge of Global Transparency, RAND Corporation and ASPRS (American Society for Photogrammetry and Remote Sensing), Santa Monica, California.

Battelle (2003), "Top Ten Strategic Technologies by 2020", *www.battelle.org*.

Bongaarts, J. and R. Bulatao (eds.) (2000), *Beyond Six Billion: Forecasting the World's Population*, National Academy Press, Washington, DC.

Boxer, B. (2002), "Global Water Management Dilemmas", *Resources*, No. 146, Winter.

Cliff, R. (2001), *The Military Potential of China's Commercial Technology*, Rand Corporation, Santa Monica, California.

Commission on the Future of the United States Aerospace Industry (2002), *Final Report of the Commission on the Future of the United States Aerospace Industry*, November, National Science and Technology Council, Arlington, Virginia

Dasgupta, S., B. Laplante, H. Wang and D. Wheeler (2002), "Confronting the Environmental Kuznets Curve", *Journal of Economic Perspectives*, Vol. 16, No. 1, Winter 2002, pp. 147-168.

Dewar, J.A. (2002), *Assumption-based Planning: A Tool for Reducing Avoidable Surprises*, Cambridge University Press, Cambridge.

Eberstadt, N. (2002), "The Future of AIDS", *Foreign Affairs*, November.

Eurospace (2004), *Facts and Figures: The European Space Industry in 2002*, Eurospace, Paris.

Fischer, S. (2003), "Globalization and its Challenges", *American Economic Review*, Vol. 93, No. 3, pp. 1-30.

Frederick, K. (2002), "Handling the Serious and Growing Threats to Our Most Renewable Resource – Water", *Resources for the Future Issue Brief* 11, Washington, DC.

Futron Corporation (2002), "Space Transportation Costs: Trends in Price per Pound to Orbit 1990-2000", Futron Corporation, Bethesda, Maryland.

Gallula, K. and P. Révillion (2002), "Satellite TV and Video Services World Survey and Prospects to 2010: A New Media Is Born (Executive Summary)", Euroconsult, Paris.

Gaubert, A. (2002), "Public Funding of Space Activities: A Case of Semantics and Misdirection", *Space Policy*, 18, pp. 287-292.

Giget, M. (ed.) (2002a), *Key Trends in Satcoms Economics*, World Summit for Satellite Financing, Euroconsult, Paris.

Giget, M. (ed.) (2002b), *Long-term Trends in Demand for Launch Services*, World Summit on the Space Transportation Business, Euroconsult, Paris.

Giget, M. (ed.) (2002c), "World Market Prospects for Public Space Programs (Executive Summary)", Euroconsult, Paris.

Gordon, B.K. (2003), "A High-risk Trade Policy", *Foreign Affairs*, July/August.

Hertzfeld, H. and M. Fouquin (2003), "Economic Conditions and the Space Sector", OECD Working Paper.

Hertzfeld, H. (2002), *Space Economic Data*, George Washington University, Washington, DC.

Institut français des relations internationales (2003), *Les grandes caractéristiques du commerce international au XXe siècle*, Paris.

Intergovernmental Panel on Climate Change (2001), *Climate Change Report 2001*, IPCC, Geneva.

International Conference on Water and the Environment (1992), "Development Issues for the 21st Century", Dublin.

International Energy Agency (2002), *Longer-term Energy and Environment Scenarios: Three Exploratory Scenarios to 2050*, OECD, Paris.

Japan National Institute for Defense Studies (1999), *1999-2000 Report on Defense and Strategic Studies*, Tokyo.

Joint Doctrine and Concepts Centre (2003), *Strategic Trends 2015*, Ministry of Defence, United Kingdom.

Kane, T. and M. Mowthorpe (2003), "The Space Sector and Geopolitical Developments", OECD Working Paper.

Kolovos, A. (2002), "Why Europe Needs Space as Part of Its Security and Defence Policy", *Space Policy*, 18, pp. 257-261.

Kupchan, C. (2002), "The End of the West", *The Atlantic Monthly*, November, No. 4, pp. 42-44.

Lewis, J. (2002), *Preserving America's Strength in Satellite Technology*, A Report of the CSIS Satellite Commission, The CSIS Press, Washington, DC.

Lutz, W., W. Sanderson and S. Scherbov (2001), "The End of World Population Growth", *Nature*, Vol. 112.

Macauley, M. and D. Chen (2003), "Space Resources and the Challenge of Energy and the Environment", OECD Working Paper.

Martell, W.C. and T. Yoshihara (2003), "Averting a Sino-US Space Race", *The Washington Quarterly*, Autumn, pp. 19-35.

National Audit Office (2000), *The Private Finance Initiative: The Contract for the Defence Fixed Telecommunications System*, Ministry of Defence, London.

National Intelligence Council (2000), *Global Trends 2015: A Dialogue About the Future with Nongovernment Experts*, National Intelligence Council, Washington, DC.

National Research Council (1996), *Meeting the Challenges of Megacities in the Developing World: A Collection of Working Papers*, National Academy Press, Washington, DC.

OECD (1998), *21st Century Technologies: Promises and Perils of a Dynamic Future*, OECD, Paris.

OECD (2004), *Space 2030: Exploring the Future of Space Applications*, OECD, Paris.

Oliker, O. and T. Charlick-Paley (2002), *Assessing Russia's Decline: Trends and Implications for the United States and the US Air Force*, Rand Corporation, Santa Monica, California.

Peeters, W. (2000), *Space Marketing: A European Perspective*, Kluwer, Dordrecht.

Peeters, W. (2002a), "Effects of Commercialisation in the European Space Sector", *Space Policy*, 18, pp. 199-204.

Peeters, W. (2002b), "Space Commercialisation Trends and Consequences for the Workforce", International Astronautical Federation Congress, Paris.

Peeters, W. and C. Jolly (2003), "Evaluation of Future Space Markets", OECD Working Paper.

Roco, M.C. and W.S. Bainbridge (eds.) (2002), *Converging Technologies for Improving Human Performance: Nanotechnology, Biotechnology, Information Technology and Cognitive Science*, World Technology Evaluation Center (WTEC) and National Science Foundation (NSF), Arlington, Virginia.

Salin, P. (2001), "Privatization and Militarization in the Space Business Environment", *Space Policy*, 17, pp. 19-26.

Salin, P. (2002), "An Overview of US Commercial Space Legislation and Policies – Present and Future", *Air and Space Law*, 27:3, pp. 209-236.

United Nations (2002), *World Population Prospects: The 2002 Revision. Highlights*, New York.

United Nations Environment Program (2002), *Global Environmental Outlook 2002-2032*, Produced by the UNEP GEO team, Division of Early Warning and Assessment (DEWA), United Nations Environment Programme, Nairobi, Kenya.

UNESCO (2003), *Water for People, Water for Life: The United Nations World Water Development Report*, Paris.

Van der Heijden, K. (1996), *Scenarios: The Art of Strategic Conversation*, Wiley, New York.

Wack, P. (1985a), "Scenarios: Shooting the Rapids", *Harvard Business Review*, November/December, pp. 139-150.

Wack, P. (1985b), "Scenarios: Uncharted Waters Ahead", *Harvard Business Review*, September/October, pp. 73-89.

Wood, J. and G. Long (2000), "Long-term World Oil Supply", US Energy Information Administration, Washington, DC, www.eia.doe.gov/emeu/plugs/plworld.html.

World Bank (2002), *2003 Global Economic Prospects and the Developing Countries*, World Bank, Washington, DC.

Zhi Dong, Li (2003), "An Econometric Study on China's Economy, Energy and Environment to the Year 2030", *Energy Policy* 31, pp. 1137-1150.

Chapter 2

Meeting Societal Challenges: How Space Might Help

> *Devoting resources to the development of space systems can only be justified from a socio-economic perspective if these systems significantly benefit society at large. This issue is explored by considering the specific contribution space might make to addressing five major societal challenges to be faced in coming decades: those related to the state of the environment, the use of natural resources, the increasing mobility of individuals and products and its consequences, growing security threats and the shift towards the information society.*
>
> *The main conclusion is that space applications have already been useful from an overall societal perspective and could be of further help in the coming decades. However, this would require fulfilling several conditions: i) further development of space systems; ii) better integration of these systems with ground-based systems; and iii) sustainable and predictable provision of space-based services.*

Introduction

From a long-term socio-economic perspective, governments' decisions to invest resources in the development of space applications – and to encourage the private sector to do so – must be based on a sound appreciation of the contribution space can make to solving the major challenges that society at large will have to face over the coming decades.

On the basis of the scenario-based exploration of the future conducted in the second phase of the project, two main types of challenges appear particularly important: those related to threats to the physical environment and the management of natural resources and those related to major trends that will shape society.

This chapter assesses how space solutions may help address some of these challenges. More specifically, five areas in which space might contribute receive special attention: the environment (climate change, pollution); the management of natural resources (water, forests, energy) and agricultural practices; the increasing mobility of people and goods worldwide and its consequences; growing security concerns throughout society; and the move to the knowledge society.

Space and the environment

Space technology can contribute to government action directed towards mitigating the effects of climate change. First, it can help improve our understanding of the complexities of climate change and ecological processes and provide valuable input for the formulation of sounder environmental policies. It can also support the effective implementation of policies aimed at reducing greenhouse gas (GHG) emissions (at national and international levels).

As noted in the second phase of the project, the environmental outlook over the coming decades is poor. If – as expected in a "business-as-usual" scenario – GHG emissions more than double over the next 30 years, temperature increases may become inevitable. In this case, a noticeable rise in sea levels, more unstable weather conditions and a shift in endemic and infectious diseases to currently temperate zones will occur. Implementing appropriate GHG abatement policies at international level will be difficult because of economies' high dependence on fossil fuels and because of the externalities involved. Higher levels of pollution are also expected in large parts of the developing world, together with further deforestation, soil erosion and reduced biodiversity.

In light of this challenge, policy circles throughout the world are giving greater attention to moving the world's economy towards a more sustainable development path. Energy-related concerns focus on the environmental effects of fossil fuel use and GHG emissions; the possible role of international agreements, such as the Kyoto Protocol, in controlling carbon emissions and in transferring energy-efficient and related technologies to developing countries; the benefits and costs of increasing the supply and use of renewable energy; and new institutional and economic regimes such as a market in which carbon emissions are traded.

To address these issues, OECD Environmental Ministers adopted in May 2001 an *OECD Environmental Strategy for the First Decade of the 21st Century*, with a view to supporting the environmental aspect of sustainable development in a cost-effective and equitable manner (OECD, 2001). The OECD Secretariat was tasked with conducting annual reviews of the progress achieved by member governments in implementing the strategy. The most recent review concluded that while OECD countries have made some headway, they are not doing enough (OECD, 2004a). In particular, current policies are insufficient to adequately protect biodiversity or address climate change, and the decoupling of environmental pressures from economic growth in key sectors is proceeding too slowly (OECD, 2004b).

To remedy these shortfalls, governments may wish to pay greater attention to the contribution that space technologies might make to the implementation of their environmental strategies. These technologies include space-based remote sensing for monitoring environmental conditions, navigation systems for monitoring and managing traffic flows, and perhaps in the more distant future, satellite solar power to supply renewable energy.

The use of space technology as an input to the formulation of environmental policies

A major difficulty for the formulation of environmental policies is insufficient understanding of complex climate change processes at regional and global levels; it is not clear which measures are likely to be effective in alleviating the harmful effects of climate change. This is a serious problem, since such policies are likely to face strong political opposition because they may impose significant costs for businesses and society at large in the shorter run.

Scientists have been able to document rapid changes in the Earth's environment over the last two centuries or so. For example, they have convincingly established that carbon dioxide levels have risen by 25% since the industrial revolution and that about 40% of the world's land surface has been transformed by human activity. However, scientists still debate the cause-and-effect relationships among the Earth's lands, oceans and atmosphere, and there are differing views on the impacts, if any, that these rapid changes will have on future climate conditions. Scientists need to make many measurements throughout the world, over a long period of time, to assemble enough information to construct models accurate enough to enable them to forecast the causes and effects of climate change. One of the most efficient ways to collect this information is through the use of space-based "remote sensors" (instruments that can measure parameters like temperature from a distance), in co-ordination with ground-based instruments.

A number of recent initiatives foster the development and use of Earth observation data. Box 2.1 summarises some initiatives at the international level.

In the United States, the National Aeronautics and Space Administration (NASA) launched in the 1990s an Earth Observing System (EOS) to initiate a systematic international study of planet Earth. The system is intended to provide a maximum amount of data for understanding climate change at world level. It has three main components: i) a series of satellites specially designed to study the complexities of global change; ii) an advanced computer network for processing, storing and distributing data (EOSDIS); and iii) teams of scientists throughout the world who study the data (NASA, 2003). The system includes three large satellites:

- The *Terra* satellite launched in 1999, is the EOS flagship. It provides global data on the state of the atmosphere, land and oceans, their interactions with solar radiation and with one another.
- The main focus of the *Aqua* satellite, launched in 2002, is the multidisciplinary study of the Earth's interrelated processes (atmosphere, oceans and land surface) and their relationship to changes in the Earth system.
- The *Aura* satellite, launched on 15 July 2004, focuses on the measurement and transformation of atmospheric trace gases. The mission is to study the chemistry and dynamics of the Earth's atmosphere from the ground through the mesosphere.

Satellites that complement EOS include those developed in co-operation with international partners, such as: the Tropical Rainfall Measuring Mission (TRMM) with the Japanese Aerospace Exploration Agency (JAXA); Jason-1, with the French space agency, Centre National d'Études Spatiales (CNES); and the Gravity Recovery and Climate Experiment (GRACE) with the German aerospace research centre (DLR, *Deutsche Forschungsanstalt für Luft- und Raumfahrt*).

Important European efforts started in 1991 with the development of ERS-1 (European Remote Sensing Satellite), primarily for ocean and ice monitoring and based mostly on all-weather radar instrumentation. It was followed in 1995 by ERS-2 (still in operation today) and in 2002 by Envisat (Environment Satellite), a very ambitious spacecraft which provides improved

> **Box 2.1. International initiatives for developing the use of satellite Earth observation data**
>
> **The Committee on Earth Observation Satellites (CEOS)** was created in 1984 as an international forum of the world's space agencies. It co-ordinates international civil space-borne Earth observation missions to ensure that they address key questions about planet Earth. The CEOS has various *ad hoc* working groups that meet regularly, including the Disaster Management Support Group (DMSG), which developed and refined recommendations for the application of satellite data to several hazard areas. The membership includes the world's governmental agencies responsible for civil Earth observation satellite programmes, agencies that receive and process data acquired remotely from space, major intergovernmental user groups and international scientific organisations.
>
> **The Group on Earth Observations (GEO)** was founded in 2003, following the World Summit on Sustainable Development held in Johannesburg in 2002 which called for strengthened co-operation and co-ordination among global observing systems, including space and non-space systems, and the first ministerial level Earth Observation Summit, held in Washington, DC, in July 2003. The GEO aims to establish broad co-ordination of global civil observing strategies by developing a ten-year plan for the implementation of a comprehensive, co-ordinated and sustained "Earth observation system of systems" for presentation at the third Earth Observation Summit in February 2005. The membership includes 47 governments and 29 participating organisations, including CEOS, diverse scientific organisations and space agencies.
>
> **The European Global Monitoring for Environment and Security (GMES) initiative** was launched in May 1998 and adopted by the European Space Agency and the European Union Councils in 2001. GMES aims to establish by 2008 an Earth observation capacity for Europe that will provide permanent access to reliable and timely information regarding the status and evolution of the Earth environment at all scales. The initiative includes space and ground-based systems. The GMES plans to provide information to improve the preparedness and response capacities of civil protection and other security-related authorities for crisis and disaster management. It is to become the European contribution to the "Global Earth observation system of systems", but will also satisfy the specific needs of European policy makers.
>
> **The Integrated Global Observing Strategy (IGOS) partnership**, co-chaired by the CEOS Chairman, seeks to reduce observation gaps and unnecessary overlaps and to harmonise research along common interests. IGOS focuses on a number of themes, including oceans, carbon and water cycles, solid Earth processes, coastal zones (including coral reefs) and geohazards. Membership includes 14 international bodies concerned with the observational component of global environmental issues, from the perspectives of both research and long-term operational programmes (*e.g.* CEOS, World Meteorological Organisation, World Climate Research Programme, United Nations Educational, Scientific and Cultural Organization, Food and Agriculture Organization of the United Nations, Global Ocean Observing System, United Nations Environment Programme).

ERS-type data and important information on the chemistry of the atmosphere in relation to ozone depletion and GHG processes. These efforts have been complemented by work undertaken in Ispra, Italy, by the Space Application Institute (SAI) to exploit remote sensing data, notably to generate crop statistics (i.e. the Monitoring Agriculture with Remote Sensing, or MARS programme) from satellite-based optical imagery. Related initiatives include the monitoring of tropical forests (i.e. the Tropical Ecosystem Environment Observations by Satellites – TREES project) and the surface of oceans as well as the generation of weekly maps of global vegetation cover indices, through the use of the vegetation instruments carried on the SPOT-4 and SPOT-5 satellites (Brachet, 2004). Europe's GMES initiative is to establish a structured European framework for data integration and information management so as to provide users with timely data, information and knowledge of good quality; it is expected to be fully operational before the end of the decade.[1]

Remote sensing from space is thus a powerful tool. Space-based observations make it possible to see the Earth as a dynamic, integrated and interactive system of land, water, atmosphere and biological processes. With space-based and ground-based instruments, one can see and feel the pulse of the entire planet from its upper atmosphere to the depths of its oceans. A broad spectrum of mapping and forecasting instruments provide the tools for witnessing deforestation, measuring tropical rainfall, assessing crop health or monitoring loss of Arctic ice cover in real time. Hurricanes, dust storms and even agricultural and industrial pollution plumes can be traced as they travel and affect the climate from continent to continent. Sea surface temperature can be measured and reveal "hot spots" where unusually high temperatures threaten the coral reefs on which more than 30 million people worldwide depend.

Earth observation also leads to better understanding of destructive climatic phenomena such as El Niño (Box 2.2). The floods, droughts and fires that accompanied El Niño in 1997-98 took more than 30 000 lives, displaced hundreds of millions of people and cost nearly USD 100 billion. Today, Earth observation from the TOPEX/Poseidon and Jason missions provides the basic data on ocean temperature and wind velocity and direction just above the ocean surface that are needed to predict El Niño events months in advance, allowing regions and countries to prepare in advance.

In addition to global climate issues, local environmental issues are also important both for OECD and non-OECD countries. Space assets increasingly allow for monitoring specific regional situations over time, and thus can contribute to forecasting possible degradation of the environment (e.g. water and ground pollution) and to planning for remedial action.

> Box 2.2. **Estimated benefits of forecasting El Niño using space data**
>
> Because Earth observation data can provide a better understanding of climate systems and phenomena such as El Nino, they can be valuable for evaluating potential disasters through better forecasting. Indicative benefits of improved El Niño forecasting in various sectors include:
>
> - Worldwide agriculture benefits of better El Niño forecasts could amount to USD 450-550 million a year.
> - Benefits to US agriculture due to altering planting decisions for El Niño, normal and La Niña years have been estimated at USD 265-300 million annually.
> - Benefits to Mexican agriculture could reach USD 10-25 million a year.
> - An analysis of NOAA's operational El Niño forecasting system, comparing its costs with anticipated benefits in the US agriculture sector alone, yields an estimated annual rate of return on that investment of 13% to 26%.
>
> Source: Based on NOAA (2004).

In addition to providing useful data on the state of the planet and climate change processes, space technology can be a useful tool for the enforcement of international environmental treaties.

The use of space technology in the implementation of greenhouse gas abatement policies

As noted in the second phase of the project, space technologies can support emissions-reduction policies, either if national governments and international agreements adopt taxes or marketable permits to enable carbon trading as an approach to control, or if they prefer binding quantitative limits on emissions. Either approach might be implemented with the support of remote sensing from space to monitor emissions and enforce compliance. Barrett (2003) argues convincingly that the foremost barrier to effective international environmental control regimes is credible monitoring and enforcement. Macauley and Brennan (1995, 2001) emphasise the potential for remote sensing from space as a monitoring and enforcement tool, given its increasingly sophisticated spectral and spatial resolution and its ability to observe activity across geographic boundaries.

Remote sensing could be useful for control programmes that monitor either the sources of fossil fuels themselves or the myriad sources of actual emissions. For instance, it could be used in efforts to control methane emissions. Scientists and policy analysts have long understood that methane is a potent greenhouse gas whose harmful effect on climate may be some 20 times greater than that of carbon dioxide, which is much more frequently discussed. Hence, along with efforts to deal with carbon dioxide, methane control may offer a cost-effective option for managing greenhouse gas.

Remote sensing could also contribute data to improve the measurement of carbon sinks (i.e. biological storage of carbon in trees, plant roots, soils, etc.) by monitoring, in co-ordination with ground-based systems, land use changes such as deforestation and reforestation. Under the Kyoto protocol (Box 2.3), the status of carbon sinks as a policy tool for GHG emissions is still to be refined; however, forestry practices can have substantial effects on the balance between stored carbon and atmospheric carbon dioxide (see Box 2.3).

Space and the management of natural resources

Space technology can contribute to more efficient use of natural resources worldwide, such as the management of water resources and forests as well as agricultural practices that are important elements of social and economic prosperity for populations around the world. As demonstrated by the conclusions of the World Summit on Sustainable Development, held in Johannesburg, South Africa, in 2002, water, energy, health, agriculture and biodiversity are elements that decision makers need to take into account in their development policies (United Nations, 2002).

Dealing with these problems will not be easy in the coming years. As illustrated by the OECD Environmental Strategy review mentioned earlier, progress is painfully slow. Given the highly charged political environment in which such policies are developed and implemented, they need a sound factual basis and effective mechanisms to monitor their application. Space technology might help in both cases.

Space and the management of energy

As noted in the second phase of this project, fossil fuels are expected to continue to dominate energy consumption over the next few decades. Oil will remain the most important fuel. Although reserves should be sufficient to meet demand over the next 30 years, major investments will be needed for exploration, extraction and transport. Moreover, the price of oil may become more volatile as conventional reserves decrease and become more concentrated in the Persian Gulf. Gas is likely to be the fuel of choice because of its relatively low carbon content, while the demand for coal should increase more slowly.

Steering the energy system away from fossil fuels will be extremely difficult, given the huge amount of resources that have been devoted to its development. It will call for a deliberate, substantial and sustained effort by

Box 2.3. **The Kyoto Protocol**

Background

Since 1988, the Intergovernmental Panel on Climate Change has reviewed scientific research and provided governments with summaries and advice on climate problems. In the late 1990s, many countries signed an international treaty – the United Nations Framework Convention on Climate Change – to begin to consider what could be done to reduce global warming and to cope with temperature increases. In 1997, governments agreed an addition to the treaty, called the Kyoto Protocol, which has more powerful (and legally binding) measures. The Protocol took effect in late 2004.

Main features

The Protocol has mandatory targets for GHG emissions for the world's leading economies that have accepted it. These targets range from a reduction of 8% to an increase of 10% of the countries' individual 1990 emissions levels "with a view to reducing their overall emissions of such gases by at least 5% below existing 1990 levels in the commitment period 2008 to 2012". The limits call for significant reductions in currently projected emissions.

Commitments under the Protocol vary from nation to nation. The overall 5% target for developed countries is to be met through cuts (from 1990 levels) of 8% in the European Union (then EU15), Switzerland, and most central and east European countries; 6% in Canada; 7% in the United States (although the United States has since withdrawn its support for the Protocol); and 6% in Hungary, Japan and Poland. New Zealand, Russia and Ukraine are to stabilise their emissions, while Norway may increase emissions by up to 1%, Australia by up to 8% (Australia subsequently withdrew its support for the Protocol), and Iceland by 10%. The EU has made its own internal agreement to meet its 8% target by distributing different rates to its member states.

Methodology: To compensate for the sting of "binding targets", as they are called, the agreement offers flexibility in how countries may meet their targets. For example, they may partially compensate for their emissions by increasing "sinks" – i.e. forests, which remove carbon dioxide from the atmosphere. That may be accomplished either on their own territories or in other countries. Or they may pay for foreign projects that result in reductions in greenhouse gases. Several innovative mechanisms have been set up for this purpose, such as "emissions trading" (Article 17 of the Kyoto Protocol).

Source: UNFCCC (2004), *http://unfccc.int*.

governments to promote the development and use of renewables. Efforts are under way in several member countries. In this regard, Europe is already the world leader in renewable energy. Denmark has some 2 300 wind turbines which supply 15% of its electricity. Germany is on course to have 140 000 solar-panelled rooftops by 2005. More than half of Scandinavia's energy comes from hydropower. In the United States, several state governments and regulatory bodies have begun to require power utility companies to use alternative energy sources for a stated percentage (usually relatively small) of their power output. The percentage may increase over time as more economically efficient technologies for alternative sources are developed.

As part of efforts to curb carbon emissions under the Kyoto Protocol, the European Commission has pledged that renewable sources will make up 22% of Europe's energy supply by the end of this decade (up from 14% in 1997). As the importance of the renewables sector grows, the idea has arisen of using satellite data to better exploit various energy sources. In fact, satellites do generate a wide variety of data that can help with many aspects of the building and management of renewable energy plants. In this regard, the following results of a workshop held in 2003 by the European Space Agency (ESA) at its Frascati centre are interesting.

Solar power. Meteorological satellites such as the Meteosat Second Generation (MSG) series can provide "sunshine maps" to help select optimal sites for new solar-cell plants. Moreover, space data can be used to help quantify the potential power from a given solar plant and its associated performance.

Wind energy. Selecting the optimal location for wind farms is very important. Satellite data on land use, surface topography and roughness can improve the accuracy of the regional wind atlases currently used to site land-based wind farms. However, offshore wind farms are likely to become the dominant form of wind farms because they are more productive than land-based ones and because land-based sites are saturated. The problem is the almost total lack of available offshore wind data. Furthermore, existing data mainly record extreme wind events, while *in situ* data gathering is costly and provides data only for a small area. Satellites enable a shift from a local to a global view. The sophisticated Synthetic Aperture Radar (SAR) instruments on board ESA's ERS-2 and Envisat satellites can provide high-resolution 100-metre data on the wind field, and a decade-long data archive is available.

Hydropower. Currently supplying around one-fifth of the world's energy needs, hydropower is a pollution-free power source that requires only the flow of water to spin a turbine. Accurate quantification of how much water will flow from a given region at any one time is extremely useful for optimising hydroelectric power production, deciding dam levels and setting electricity prices. In Norway for instance, where hydropower supplies almost all of the country's electricity needs, around half of all winter precipitation accumulates on the ground throughout the winter as snow. By measuring snow coverage and thickness, and then combining this information with meteorological data such as ground temperature, run-off can be accurately modelled and predicted. Together with ground observation, optical Earth observation data are already used to monitor snow coverage, although clouds severely limit the operational use of optical data. However, radar instruments such as those aboard ERS-2 and Envisat have the potential to greatly supplement the amount of snow data gathered, because they can measure through the clouds.[2]

Power from space. When looking at possible long-term energy solutions, several countries' experiments with satellite laser technologies and research on space-generated power may provide some interesting prospects.[3] Recent work indicates that the collection and transmission of power from space could become an economically viable means of exploiting solar power within the next couple of decades (David, 2003). However, technological advances are needed to allow space-generated power to compete with current Earth-based alternatives. The US National Research Council (2001) argues that the ultimate success of the terrestrial application of powering-beaming satellites will critically depend on "dramatic reductions" in the cost of transport from Earth to geosynchronous orbit.

Space and water management

Concerns related to water have focused on the supply, distribution and quality of water. Although water itself is not universally scarce, its quality and its distribution to regions and communities where supplies are limited remain highly problematic. Water is now among the highest international priorities for addressing natural resources issues.

Space-related innovations include adapting existing Earth-systems modelling data for assessments of water resources (for example, the effects of greenhouse warming on regional water resources) and the diffusion of tools such as geographic information systems (GIS) and global positioning system (GPS) devices. These tools can help in the collection and interpretation of the data necessary to develop more integrative models. Remote sensing can also monitor water distribution, multi-use reservoir management, and the reservoir compensation flow releases that take place across jurisdictions within and among countries, since the boundaries of watersheds rarely coincide with administrative boundaries. Space technology has also proved useful for detecting water-based diseases, notably the presence of cholera, a disease that remains a threat for a large portion of the world's population.

Remote sensing can also help alleviate the economic and population loss associated with water-related natural disasters. Some 90% of all natural disasters are hydrometeorological, and are caused by hurricanes and floods

(United Nations, 2003). Economic losses from natural catastrophes are primarily due to these events and have increased markedly in the past decade, largely because of increases in shoreline property values and population growth along coastlines around the world.

Space and forestry management

Concerns related to deforestation have focused on its consequences for the overall level of greenhouse gases worldwide, its effect on local climate and hydrology, as well as its adverse impact on biodiversity. First, the loss of forests has a profound effect on the global carbon cycle. From 1850 to 1990, deforestation worldwide caused the release of 122 billion metric tons of carbon dioxide into the atmosphere, and the current rate is approximately 1.6 billion metric tons a year. As the burning of fossil fuels (coal, oil, gas) releases about 6 billion metric tons a year, deforestation clearly makes a significant contribution to the increase of carbon dioxide into the atmosphere. Second, tropical deforestation also affects the local climate by reducing evaporative cooling from both soil and plant life. As trees and plants are cleared away, the moist canopy of the tropical rain forest quickly diminishes and temperatures rise. Third, deforestation has an adverse impact on biodiversity. Worldwide, the biodiversity of planet Earth includes five to 80 million species of plants and animals. Tropical rain forests cover only 7% of the total dry surface of the Earth but hold over half of all these species. Every day, as the tropical rain forests are cleared, species disappear. The exact rate of extinction is not known, but estimates indicate that up to 137 species disappear worldwide each day.

While it is impossible to measure the cost to society that is likely to result from the loss of biodiversity, it is clear to most experts that the rapid extinction of species currently under way is bound to have dire consequences for the survival of humanity in coming decades. On a more modest level, estimates of the economic cost of deforestation can be made. For instance, according to the European Commission, each hectare of forest lost to fire costs Europe's economy between EUR 1 000 and EUR 5 000.

Space technology is useful for managing forest resources more effectively and combating deforestation. The use and management of forest resources need to be based on the mapping and inventory of the forestry environment. In addition, the changing state of the forest, as a result of natural causes or human activity (felling, clearing, fire, reforestation, decline, regeneration, etc.), needs to be monitored. Remote sensing and geographic information systems provide for the continuous monitoring of forest developments by detecting changes, and the findings can be integrated into existing databases.

High-resolution satellite imagery is particularly useful for investigating and monitoring forest resources. Compared with information acquired by traditional methods, these data offer certain advantages. First, satellite imagery can cover vast expanses of land (thousands to tens of thousands of square kilometres on one image), cover the same area regularly and record the information in different wavelengths, thereby tracking the state of forest resources. Second, satellite data can be acquired without encountering administrative restrictions.

Combined with *in situ* data, satellite images taken on a regular basis provide forest managers and developers with:

- A characterisation and description of the physical environment of the forest.
- A cartographic representation of forest change and a corresponding statistical inventory in terms of felling and reforestation over time. This helps track management progress and spotlight illicit felling, making it easier and cheaper to combat illegal logging.
- The development and structuring of a database from the mapped changes and available conventional data, in the form of a GIS. This facilitates the regular monitoring and continuous management of the forest.

Satellite imaging can also be usefully applied for the monitoring of mangrove forests. Such forests are one of the world's most important coastal ecosystems in terms of primary production and coastal protection.

Finally, wetlands mapping is needed to better understand various wetlands conditions and to delineate the aerial extent and boundaries of wetlands and, in particular, losses in coastal and inland wetlands. These maps can serve as baseline data for classifying coastal zones into preservation, conservation and development zones. Remote sensing data can provide useful information about the aerial extent, conditions and boundaries of wetlands and have proven extremely useful for wetlands mapping and for determining high and low water lines.

Space and agriculture

The impact of the widespread adoption of modern agricultural practices, which accelerated after World War II, is also a major area of environmental concern. In the United States, for instance, agricultural productivity changed more rapidly between 1950 and 1975 than at any other time in American history. Although the acreage farmed dropped by 6% and the hours of farm labour decreased by 60%, farm production per hour of on-farm labour practically tripled, and total farm output increased by more than half. These dramatic changes were due to technological innovations, development of hybrid strains and other genetic improvements, and a fourfold increase in the use of pesticides and fertilisers. The result of all these changes has been that agriculture has become more intensive, producing higher yields per acre through greater use of chemicals and technological inputs. It also has had a number of potentially detrimental environmental consequences, ranging from rapid erosion of fertile topsoil to contamination of drinking water supplies by the chemicals used to enhance farmland productivity.

Precision agriculture. Two relatively recent developments could bring the fruits of the information revolution to farming. The first is precision farming, or the use of detailed data for agronomic assessment, and the second is automated control, or the use of automated machines to apply agronomic treatments. The combination of these two developments may improve the overall efficiency of farming and reduce the environmental impact of crop production.

Before the industrialisation of agriculture, all farming was, by default, precision agriculture. "Farming by foot" meant that the farmer had detailed knowledge of the conditions within each field. Increased mechanisation and larger fields made it more difficult to be aware of variability within fields. Technological advances in other areas have led to the application of information technology to farming. This is facilitated by technologies such as global navigation space systems (GNSS) and spaced-based augmentation systems (SBAS), geographic information systems, miniaturised computer components, automatic control and in-field and remote sensing. Where farmers once relied on empirical observations, new sensors mounted on tractors, airplanes and satellites make systematic data available for their use.

Recent innovations in geographic information systems for urban planning have tackled the management of large databases and are now finding application in farming. Data from the field are applied to other maps of the field, such as soil composition or stalk height. These various "transient data" are then used to appraise the field and to identify areas requiring treatment. When and where these treatments are needed becomes a matter of computer algorithms and farmers' experience (Earl, 2000). Applying a planned regimen of such treatments to specific parts of large fields becomes feasible thanks to improvements in automated control and positioning. For example, GPS allows for resolution within meters, and high-precision differential GPS (DGPS) can accurately target a moving vehicle within 30 cm. Such precision can make targeted treatments possible and can also be used as a field-scouting tool.[4]

The developing world, particularly China, may realise the greatest gains from precision agriculture (Zhang et al., 2002). As a rapidly modernising economy, China is going through what industrialised countries experienced a century ago, but at a much more rapid pace. There is a drain on its traditional land-scarce and labour-abundant agriculture as migrant workers leave rural areas for urban opportunities. Some observers predict a step-by-step diffusion of precision agriculture technologies into Chinese agriculture, first with experimental and project farms, then with pre-assembled modules or technology kits that can be distributed easily to farmers.

Space and the mobility challenge

The world will face a major mobility challenge. On the one hand, mobility is essential to modern civilisation and to meeting human needs; it facilitates economic activities and economic relations and literally makes modern economies possible, as cities would not otherwise exist. On the other hand, it is increasingly realised that the world's continuing and growing demand for mobility cannot be met simply by expanding today's means of transport.

Mobility has increased continuously over the last few decades. Particularly significant has been the growth in air and car travel (Figure 2.1). For instance, between 1965 and 1989, passenger kilometres travelled by air and car grew by 700% and 287%, respectively, in western Europe, whereas travel by train and bus grew only by 33% to 38%) (Nijkamp et al., 1998).

Figure 2.1. **Transport growth by mode in western Europe**[1] **since 1965**
Index: Million passenger km

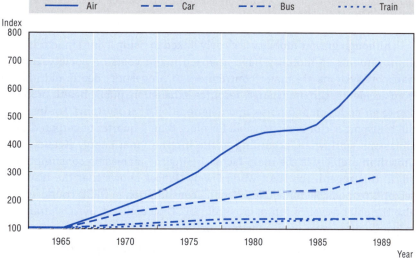

1. Excluding Luxembourg, Greece and Ireland.

Source: Nijkamp, P. et al. (1998), Transport Planning and the Future, John Wiley and Sons Ltd., London.

Air travel has grown rapidly in other regions as well (Figure 2.2). According to the International Civil Aviation Organization (ICAO), the volume of air travel worldwide was 1 189 billion passenger kilometres in 1983 and reached 2 628 billion in 1998. Between those years, faster growth was recorded in the Asia-Pacific area (230%) and North America (130%) than in Europe (80%) (ICAO, 2001).

2. MEETING SOCIETAL CHALLENGES: HOW SPACE MIGHT HELP

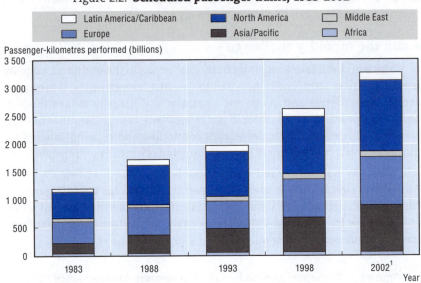

Figure 2.2. **Scheduled passenger traffic, 1983-2002**

1. Estimated.
Source: ICAO: International Civil Aviation Organization (2001), "Outlook for Air Transport to the Year 2010", Circular 281, June.

Although greater mobility is closely linked to rising living standards, past improvements in mobility have imposed major costs on society. Indeed, transport systems are major contributors to congestion, death and injuries from accidents, climate change, resource exhaustion, public health problems due to air pollution and noise, and deterioration of ecosystems. For instance, the transport sector is responsible for more than a quarter of carbon dioxide emissions worldwide, according to the International Energy Agency (IEA). Growth in mobility is the main factor driving the increasing consumption of oil. In this regard, the IEA's *World Energy Outlook* shows that the transport sector is responsible for about 55% of oil consumption in OECD countries. It is also the chief driver of future growth in OECD oil demand. By 2030, transport may account for 65% of oil consumption (IEA, 2002).

The costs of these external effects are high and vary among transport modes. For instance, the European Commission has estimated the external cost of travel, per thousand passenger kilometres, at EUR 87 for cars, EUR 48 for air, EUR 38 for bus and EUR 20 for rail (EC, 2001). The estimates do not include congestion costs.

The two modes of travel with the most rapid growth, air and car, are also those with the highest external costs. It is obvious that this trend is not ecologically sustainable, even if technological developments help to alleviate the consequences in the future. Yet, from the social and economic point of view, air and car travel has many benefits. The most obvious is speed, a highly valued quality.

The rise of the automobile reflects its advantage for the traveller over other modes of transport, notably public transport. Automobiles offer greater flexibility in terms of schedules and choice of routes and destination; this is particularly important for leisure travel, which represents 40-50% of all mileage travelled in western economies. They offer benefits in terms of travel time, travel comfort, amenities and also status and prestige that are not entirely related to "functional" mobility, when the externalities imposed on society are not taken into account.

When looking to the future, it is clear that current trends are socially and environmentally unsustainable. An analysis conducted by the World Business Council for Sustainable Development (WBCSD) is instructive in this regard. The projections are based on "business as usual assumptions": i) the mainstream projections of economic and population growth are accurate; ii) the general trajectory of technological development and its incorporation into transport systems and services continue much as over the past several decades; and iii) policies currently in place continue to be implemented, but no major new initiatives are launched (WBCSD, 2004).

The results show that:

- Personal transport activities could more than double over the next three decades from about 32 trillion passenger kilometres a year in 2000 to close to 75 trillion in 2030 (Figure 2.3). Although growth is expected to be fastest in developing countries – notably in China (3% a year) and Latin America (2.9% a year) – this will not overcome the "mobility opportunities divide" between rich and poor countries and between rich and poor within countries.

- Rail and road freight transport activities should rise in the same proportions, from roughly 15 trillion tonne-kilometres a year in 2000 to 30 trillion in 2030 (Figure 2.4). India, China and other parts of Asia represent the strongest growth, totalling 12% of the average annual rate for the 2000-30 period.

- Transport-related emissions of carbon dioxide increase in the same proportion, as progress in the technical efficiency of vehicles is more than offset by the increase in the number and size of vehicles and average vehicle utilisation.

- Deaths related to road vehicles decline in OECD countries and in some "upper middle class" developing countries, but rise for at least another couple of decades in the rest of the world, and transport-related security remains a serious concern.

2. MEETING SOCIETAL CHALLENGES: HOW SPACE MIGHT HELP

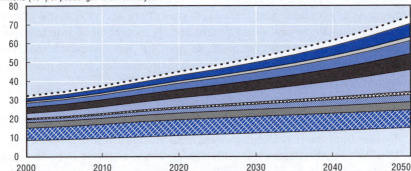

Figure 2.3. **Personal transport activity by region, 2000-50**

	Average annual growth rates	
	2000-2030	2000-2050
Total	1.6%	1.7%
Africa	1.9%	2.1%
Latin America	2.8%	2.9%
Middle East	1.9%	1.8%
India	2.1%	2.3%
Other Asia	1.7%	1.9%
China	3.0%	3.0%
Eastern Europe	1.6%	1.8%
Former Soviet Union	2.2%	2.0%
OECD Pacific	0.7%	0.7%
OECD Europe	1.0%	0.8%
OECD North America	1.2%	1.1%

Source: WBCSD: World Business Council for Sustainable Development (2004), "The Sustainable Mobility Project, Mobility 2030: Meeting the challenges to sustainability", Overview 2004, July.

- Congestion increases in all major urbanised areas in both the developed and the developing world. This adversely affects the reliability of personal and goods mobility.
- Transport's resource "footprint" grows as transport-related use of materials, land and energy increases.

Clearly, "business as usual" will not do. Today, only about 12% of the world's people are motorised. If the benefits of mobility are going to be available for a larger segment of the globe's population over the longer term, the challenges associated with current means of transport have to be addressed effectively. In short, transport systems have to become more efficient, more equitable, more technologically advanced, and less environmentally and socially disruptive, while preserving the attributes that make mobility desirable.

Figure 2.4. **Road and rail freight transport activity by region, 2000-50**

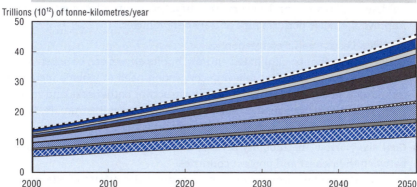

	Average annual growth rates	
	2000-2030	2000-2050
Total	2.5%	2.3%
Africa	3.4%	3.1%
Latin America	3.1%	2.8%
Middle East	2.8%	2.4%
India	4.2%	3.8%
Other Asia	4.1%	3.7%
China	3.7%	3.3%
Eastern Europe	2.7%	2.8%
Former Soviet Union	2.3%	2.2%
OECD Pacific	1.8%	1.6%
OECD Europe	1.9%	1.5%
OECD North America	1.9%	1.7%

Trillions (10^{12}) of tonne-kilometres/year

Source: WBCSD: World Business Council for Sustainable Development (2004), "The Sustainable Mobility Project, Mobility 2030: Meeting the challenges to sustainability", Overview 2004, July.

In this regard, the 2004 WBCSD report outlines seven sustainable goals established by the Sustainable Mobility Project (SMP), which involved 12 companies in the energy and automobile manufacturing sectors.[5]

- **Goal one:** Reduce conventional emissions from transport so that they do not constitute a significant public health concern anywhere in the world.
  Technology can drive conventional emissions down in developing countries. In the developed world, the focus will shift from setting standards towards making sure prescribed emission levels are met. "High emitter" vehicles are likely to be the focus of attention.

- **Goal two:** Reduce GHG emissions from transport to sustainable levels.
  The SMP members argue that society's long-term goal should be to eliminate transport as a major source of GHG emissions but warn that this cannot be achieved until well beyond 2030. In addition, they believe that the

portfolio of technology options they are currently pursuing for vehicle fuels and powertrains will be a significant factor in achieving stabilisation of CO_2 emissions. Stabilising CO_2 emissions from transport should be part of a larger array of policy measures aimed at reducing GHG emissions from all major sources.

- **Goal three:** Reduce significantly the number of transport-related deaths and injuries worldwide.
Programmes to reduce deaths and serious injuries from road vehicle crashes should focus on at least four factors: driver behaviour, improvements in infrastructure, better technologies for crash avoidance and injury mitigation.
- **Goal four:** Reduce transport-related noise.
Overall, traffic noise is not likely to decrease. However, local authorities can address the increase in traffic noise through a combination of road surfaces and barriers that dampen noise and by restricting the modification of vehicles by owners and others. Manufacturers are also continuing to improve the noise performance of transport vehicles.
- **Goal five:** Lessen traffic congestion.
Congestion cannot be eliminated entirely, but its effects can be substantially lessened. Actions aimed at relieving congestion need to include increasing infrastructure capacity, eliminating infrastructure bottlenecks and making more efficient use of existing transport systems and infrastructure. Pricing strategies could play an important role in this regard.
- **Goal six:** Narrow the mobility divides that exist within all countries and between the richest and poorest countries.
This divide inhibits growth and works against the efforts of the poorest countries and peoples to escape poverty. Sustainable mobility requires narrowing the gap.
- **Goal seven:** Improve mobility opportunities for the general population in developed and developing societies.
Improved mobility opportunities for all societies – developed and developing – is an important prerequisite for future economic growth and a basis of a more sustainable global transport system based on wider access and greater affordability.

Space-based solutions, notably the use of global navigation satellite systems and satellite telecommunications, may increasingly help meet the mobility challenges.[6] Indeed, the ability to determine accurately and communicate one's position at any moment thanks to GNSS is starting to have a major impact on the management of ship and lorry fleets, road and rail traffic monitoring, the mobilisation of emergency services, the tracking of goods carried by multimodal transport and air traffic control. Communications satellites are also having a marked impact on travellers' connection to the world, especially when they are in remote areas where no terrestrial communications networks are available.

Road transport. For road transport, space-derived navigation services can play a role in reducing vehicle miles travelled by improving routing and delivery for both fleet vehicles and passenger autos. Personalised radio and multimedia services to cars, with nearly real-time services, can be provided by existing low Earth orbit space systems, and soon by the new Ku-band satellite using existing geostationary satellites. SES Global, for instance, is working with ESA and other partners (*e.g.* DLR, BMW, Dornier) to develop such innovative systems.[7]

Relatively innovative demand management strategies can help to curb the growth in traffic. Useful tools for these purposes include:

- Smart cards for collecting fees to access toll roads.
- Development of real-time passenger information systems for managing traffic flows, pinpointing locations even when a driver is injured or cannot describe his or her whereabouts, and automatic vehicle identification using GNSS.
- Dynamic scheduling and routing.
- Intelligent transport systems (ITS) technology to improve the operation and efficiency of existing highways. It includes sensing and communications technologies, traveller information systems, payment mechanisms and traffic management.[8]

Air traffic. Air traffic control (ATC) is another major area of application for GNSS (ICAO, 2004). The performance of any navigation system is judged by its accuracy, availability, continuity and integrity. If the performance of satellite navigation is satisfactory in these respects, it may give air traffic management increased utility, effectiveness, efficiency and flexibility.

For the aviation community to take full advantage of satellite technology in oceanic, en route and terminal environments, the US Federal Aviation Administration (FAA), has undertaken the task of integrating satellite navigation into its ATC system. The Wide Area Augmentation System (WAAS), a derivative application of GNSS, provides coverage over the Americas and transoceanic routes to improve safety of life, accuracy, availability and integrity from takeoff to precision approaches. Other modes of transport also benefit from the increased accuracy, availability and integrity that WAAS delivers. The WAAS broadcast message improves GPS signal accuracy from 20 meters or better horizontal accuracy worldwide to approximately one to two meters and two to three meters vertical accuracy throughout most of the continental United States and portions of Alaska (FAA, 2004).

The benefits to civil aviation of WAAS and other regional systems, such as the European Geostationary Navigation Overlay System, the first component of the future Galileo constellation, will be substantial. These systems should improve the efficiency of aviation operations owing to:

- Greater runway capability.

- Reduced separation standards which allow increased capacity in a given airspace without increased risk.
- More direct flight paths.
- New precision approach services.
- Reduced and simplified equipment on board aircraft.
- Significant government cost savings owing to the elimination of maintenance costs associated with older, more expensive ground-based navigation aids.

One crucial reason for using regional satellite-based augmentation systems for air traffic control is their ability to meet stringent integrity monitoring requirements. In safety-of-life applications such as guiding an aircraft to a runway, integrity monitoring is essential. It deals with the all-important question: "Is the error in an estimate of a given position less than a certain preset threshold?" The main objective of integrity monitoring is to warn the pilot when the navigational guidance provided may be erroneous to the point of being hazardous. The question must be answered in real time, and the consequences of a wrong answer can be severe.

Up to now the Federal Aviation Administration's WAAS and Local Area Augmentation System (LAAS) have not had an easy time with integrity monitoring for precision approaches. A recently proposed possible solution to this problem would require all users to have the richly redundant measurement sets typical of constellations of 40-plus satellites, each equipped with the protected frequency spectrum authorised by the Aeronautical Radio Navigation Service for air traffic control safety-of-life applications (Mistra et al., 2004). This will be the case when Galileo becomes operational, especially when used together with GPS. More developed GPS systems could in this way reinforce Galileo's safety-of-life applications.[9]

Space and security

Several of the major drivers of change identified in the second phase of the project and in recent studies conducted in the context of the OECD's International Futures Programme, are likely to increase the level of risk faced by our societies in future.[10] As noted in Box 2.4, such risks range from political risks (e.g. terrorism), economic risks (e.g. major economic crises) and demographic risks (related to rapid population growth and massive migration to the cities), to environmental risks, risks related to the growing mobility of persons and goods (e.g. threats to the supply chain) and technology risks (related to the vulnerability of complex ubiquitous infrastructure).

Growing concerns about risk fuel rising demand for security. This in turn causes a substantial increase in the security sector of the economy and fosters the development and implementation of security technologies, including

> Box 2.4. **Major risks facing society in coming decades**
>
> *Political risks:* While armed conflicts between nations are likely to decline in future, civil wars are expected to be on the rise. Moreover, organised crime and international terrorism may become more active in a more open world. This growing threat will be exacerbated by the proliferation of weapons of mass effects (WME).
>
> *Economic risks:* Future economic risks are also significant and may contribute to a climate of social tension, civil disobedience and acts of violence. First, globalisation will further constrain government economic policy and severely penalise poor governance. The poorest states will be the most vulnerable. Defaults could cause major hardship for populations, triggering a violent reaction against the "culprits". Second, income inequalities in developing countries are likely to increase. This may be a source of serious conflict between the elite and the rest of society and between rich and poor regions. Third, economic crises can have devastating effects on whole regions, as such crises spread ever more quickly in an increasingly interdependent world, where capital is very mobile and information on national policies is more readily available.
>
> *Demographic risks:* In slow-growing developing countries that go through a demographic transition as a result of declining birth rates, discontent may become widespread if employment growth falls short of the expected large increase in the working age population. Moreover, migration from the countryside to cities as well as from least developed countries (LDCs) to other developing and to developed countries may be a source of major social disruptions and could be a serious source of tensions and conflicts.
>
> *Environmental risks:* Global warming will result in more frequent natural hazards such as floods and fires. Population growth will also put additional pressure on the natural environment, leading to losses in biodiversity with potentially catastrophic consequences for life itself, and increasing scarcity of water in some regions may become an important source of conflict.
>
> *Mobility risks:* Growth in air, rail, road and maritime transport of goods and people increases the risk of security breaches that facilitate robbery and terrorism. Greater mobility weakens countries' ability to impede clandestine threats, and communications and supply chains that span the globe give rise to greater vulnerability.
>
> *Technology risks:* Technological advances are likely to make systems more complex and their vulnerability more difficult to ascertain; because of their interconnection, catastrophic chain reactions may occur if one of the components fails.

space-based solutions. Surveillance, tracking and authentication technologies are already widely used, and more – and more sophisticated – innovations are in the pipeline and at the laboratory stage. The future of the security sector will depend on how demand and supply developments come together and what actions are taken by governments.

The role that space technology can play

Space-based technology has three distinct characteristics that make it particularly attractive for security purposes: It can:

- Communicate anywhere in the world whatever the state of the ground-based network.
- Observe any spot on Earth very accurately and in a broad spectrum of frequencies.
- Locate, at an increasing level of precision, a fixed or moving object anywhere on the surface of the globe.

The following discussion briefly reviews how space-based technologies can help to enhance security in three areas of application: disaster and relief management; treaty monitoring; and the monitoring of hazardous goods and pollution.

Disaster relief and prevention

Satellites can provide support for disaster management throughout the disaster management cycle. First, satellite communications provide invaluable support to relief operations when terrestrial networks are down. In such cases, satellite-based solutions that support telemedicine or telehealth services may be the only ones able to function effectively. Moreover, satellites can be rapidly deployed over a crisis location or an area lacking ground infrastructure.

Second, satellite imagery can be used to estimate the scope of a disaster and the level of relief effort needed to cope with the situation as well as the way in which the effort should be deployed. An interesting example in this regard is the International Charter for Space and Major Disasters, signed on 20 October 2000. It is a joint effort by six space agencies to put space technology at the disposal of rescue authorities in the event of major disasters. Other systems under development that could be used in times of disaster include the SIASGE constellation (*Sistema Italo-Argentino de Satélites para la Gestión de Emergencias*), which will provide radar information in cases of disaster. The constellation will include Argentina's first radar satellite (SAOCOM-1), to be launched in 2005, and the COSMO-Skymed satellite, also to be launched in 2005 by the Italian space agency.

First responders need to have a good knowledge of the physical environment in which a disaster has occurred. Satellites can help by providing geospatial data that can facilitate rescue efforts, but to be useful, these data need to be easily accessible and integrated in larger information systems. The development of such a capability is currently under way in different countries, including the United States. Under a US Department of Homeland Security contract awarded in May 2004, Northrop Grumman Information Technology is to develop software that integrates existing communications and data-display systems to create a field capability that is ready for operational use. The integrated system is to give fire fighters, police and other first responders on the scene access via laptop computers to geospatial information that is currently available only at headquarters facilities. Under the first year of the contract, the company is expected to produce operationally ready field systems tailored for delivering geospatial data to ports and border patrol units, as well as a command centre for storing and distributing the information. The port security system will be demonstrated in Miami, where officials want to integrate information from such diverse sources as satellites, security cameras and aerial vehicles. Several demonstration projects targeted at first responders' uses (*e.g.* updatable geospatial information accessible via small receivers) are being conducted in different countries.

Another space-based humanitarian programme that has proved its worth over the years is the Cospas-Sarsat system, which was set up in 1982. The programme is the result of international collaboration by the United States, France, Russia and Canada. It provides an alert and satellite positioning aid function for the search and rescue of persons in physical distress, on land or at sea anywhere in the world. The system uses instruments on board eight satellites in geostationary and low-altitude Earth orbits that detect the signals transmitted by distress radio beacons and ground receiving stations that receive and process the satellite downlink signal to generate distress alerts to rescue centres throughout the world. As of December 2003, the Cospas-Sarsat System had provided assistance in rescuing more than 17 000 persons in over 4 800 incidents (*www.cospas-sarsat.org*).

Treaty monitoring

Satellites are also used in another important security area: treaty monitoring. The diversity of available instruments and the increasing amount of obtainable "open source" cross-referenceable data (notably via commercial firms) create new opportunities for using satellites (Hettling, 2003).

An important application in this regard is the Comprehensive Nuclear Test Ban Treaty (CTBT). The purpose of the CTBT, adopted by the United Nations General Assembly on 10 September 1996, is to ensure that the international community rids the world of the testing of nuclear weapons. Under Article 1 of the CTBT, each state signatory "undertakes not to carry out nuclear weapons test explosions and all other nuclear explosions, and to prohibit and prevent any nuclear explosion at any place under its jurisdiction or control". The CTBT seeks to constrain the development and qualitative improvement of nuclear weapons as well as the development of new advanced types of nuclear weapons. To monitor state parties' compliance with the treaty's provisions, an international monitoring system consisting of 321 monitoring stations and 16 laboratories in 91 countries has been set up or is in the implementation process. The monitoring stations send data in near real time to an international data centre in Vienna over a global communications infrastructure incorporating ten geostationary satellites and three satellites in inclined orbits. The global communications infrastructure will eventually support 250 thin-route, very small aperture terminal (VSAT) satellite links to the monitoring stations. This is the first global integrated satellite communications network based on VSAT technology.

Satellites have also played a major role in raising awareness of the state of the environment, leading sometimes to the implementation of successful environmental treaties. This is the case for instance for NASA's Nimbus-7 satellite. Following the discovery of the depletion of the ozone layer by a ground-based research team in 1984, the depletion was subsequently confirmed by data from sensors aboard the Nimbus-7. Imagery from the Nimbus-7 Total Ozone Mapping Spectrometer (TOMS) instrument was used to document seasonal depletion of ozone over the Antarctic, and the media's use of the images led to public awareness and a call for action. These events paved the way for the Montreal Protocol on Substances that Deplete the Ozone Layer, widely considered one of the most effective environmental treaties. Images from TOMS have given eloquent support to other scientific evidence that was crucial in expanding the protocol.

Monitoring of hazardous goods transport and polluters

The ubiquitous surveillance capability of satellites can also be applied to the monitoring of international borders as well as to the movement of ships on the high seas. For instance, in Australia, Customs Coastwatch, which manages the security and integrity of Australia's borders, has installed a world-first, high-speed satellite communications system. The Inmarsat-based technology provides high-speed transmission of real-time data and imagery between surveillance aircraft and the Coastwatch National Surveillance Centre for distribution to relevant government agencies. It is the first time the technology has been employed in civil maritime surveillance anywhere in the world.

Satellites may also be used in future to spot suspicious ships on the high seas and monitor their cargo. The possibility for terrorists to sneak dirty bombs or pathogens into a country aboard ships or in shipping containers is a major concern of security authorities. To deal with this threat, one would want ideally to be able to spot the suspicious ships or cargo while they are well out to sea, where terrorist plots could do little harm. A possible solution, soon to be tested by the US Coast Guard, is to use satellites with ship identification relays that could pick up the identification information that commercial ships broadcast today at 30-second intervals to one another or to ports, as part of an international network called the Automatic Vessel Identification System. If the test is successful, this could be a cost-effective solution in that it would not require the installation of new shipboard equipment. Having ships actively transmit the information to existing satellites such as Immarsat maritime satellites would require a new international agreement and would be costly.

A complementary approach would be to set up a satellite-based container monitoring system. This would involve the installation of electronic seals on the doors of shipping containers. If intruders avoided the seals by cutting through the tops or sides of the containers, carbon dioxide or light sensors might detect the intrusion. The container monitors would detect tampering and transmit the data wirelessly to a satellite transmitter aboard ship. The ship would transmit the data to a satellite system which would relay the information to US authorities. Tests of the container concept are under way using actual cargo containers on commercial ships.

Satellites may also be used in future to detect ship pollution. In this regard, the UK Maritime and Coastguard Agency has recently unveiled the first results of an international tripartite surveillance programme using satellites currently under way in the North Sea. The trial involves several member states of the European Union, including Germany, the Netherlands and the United Kingdom. One aim of the project is to exploit the use of new satellite technology to identify marine pollution originating from shipping. Another is to provide a deterrent to shipowners and operators. While satellite trials linked to oil spill detection are not new, this latest demonstration project is believed to be very much on an operational footing. ESA's Envisat satellite and the Canadian Radarsat satellite take several images which are then acquired and processed by the Konigsberg Satellite Services (KSAT) ground station based in Tromso, Norway.

Space and the information economy

OECD economies are increasingly based on the production, distribution and use of information and knowledge. This is reflected in the growth of high-technology investment and high-technology industries, and in the growing role of highly skilled labour in the production of goods and services.

Recent analytical and empirical economic work conducted at the OECD and elsewhere confirms the strategic importance of the creation and diffusion of information and knowledge for economic development – in developed and developing countries alike – and the key role played by education (Box 2.5).

Box 2.5. **Knowledge and economic development**

Although knowledge has always been a factor in economic development, its role in driving productivity and growth was not well understood in the past. Recent analytical and empirical economic work on "new growth theory" or "endogenous growth theory" has shed further light on this issue and has stressed in particular the key role played by investment in research and development (R&D), education and training, and the development of new managerial work structures.

Research has also shown that diffusion of knowledge via formal and informal networks is an essential element of economic performance. Knowledge is increasingly codified and transmitted through information and communications technologies (ICT). However, effective use of codified knowledge requires tacit knowledge, which includes the skills needed to use and adapt codified knowledge, and thus continuous learning by individuals and by firms.

Employment in the information economy is characterised by increasing demand for more highly skilled workers. Changes in technology, notably the development of information technologies, are making educated and skilled labour more valuable and unskilled labour less so. This means that governments must make special efforts to promote skills upgrading and to provide the enabling conditions for organisational change at firm level to maximise the benefits of technology for productivity.

The importance of the creation and diffusion of knowledge through R&D activities and education is confirmed in a recent OECD report that presents the results of an in-depth reflection on what has been driving economic growth in OECD countries over the most recent decades (OECD, 2003b). The report stresses notably the high return to investment in education, the marked positive effect of business R&D on growth and the major role played by information and communications technologies (ICT) in recent decades.

Recent World Bank studies confirm these results. They illustrate the huge payoffs that investments in knowledge can bring. For instance, through investment in education and information technologies, Ireland has transformed its once rural economy and is now Europe's largest exporter of computer software. Thirty years ago, 70% of Finland's exports were wood and paper products. Now, more than 50% of the country's exports are knowledge-intensive products and it is a world leader in information technologies.

The role that space technology can play

Space-related activities contribute both to the creation and to the diffusion of information and knowledge. It is obvious that they create new knowledge, given the amount of R&D needed to develop space systems, and the fact that a number of such activities are devoted to scientific missions. For instance, as noted above, Earth observation has generated a wealth of information on the state of the planet and on the various parameters that affect climate. Increased understanding of natural phenomena and of the impact of anthropogenic activity should provide the basis for sounder environmental policies. As well, the application of Earth observation to agriculture gives farmers new knowledge that they can use to adopt more effective agricultural practices. Even when the research effort does not directly serve economic and social objectives, it can have a major impact on society. For instance, space-based navigation systems, which were originally developed for military purposes, are finding an increasingly broad range of civil applications.

Space-based activities can also improve access to knowledge, as in the case of satellite communications, which both complement and compete with terrestrial communication systems. Digital television, third-generation mobile communications and the Internet are examples of useful platforms for the deployment of such services to which space systems can contribute.

Direct broadcast satellite (DBS) is a relatively recent development in television distribution.[11] DBS uses special high-powered Ku-band satellites that send digitally compressed television and audio signals to fixed satellite dishes. DBS services, which are part of the larger family of satellite direct-to-home (DTH) services, offer many advantages over traditional analog services, such as analog cable television. They generally offer better picture quality, more channels and additional features such as an on-screen guide, digital video recorder functionality, high-definition television (HDTV), and pay per view.[12]

In competitive markets, satellite solutions offer consumers and businesses an alternative to terrestrial technologies, thereby stimulating innovation and contributing to improving the quality of services. Satellite communications can also facilitate social inclusion by serving the needs of those who live beyond the reach of terrestrial networks. Moreover, satellite communications will be increasingly attractive to individuals in an ever more mobile society, by offering services everywhere and at all times. Business people and individual citizens will want their mobile phones to work everywhere, regardless of geographical location or network provider, so that there may well be growing demand for mobile satellite phones that give users the best of both worlds: coverage in areas covered by terrestrial technologies and the convenience of satellite phone communications in areas outside the reach of mobile phone masts.

At European level, it has been argued that satellite communications might facilitate the integration of member states and make it possible to improve the quality of services to citizens, companies and public authorities more rapidly by reinforcing the communications infrastructure linking the new members with the rest of the Union, and by facilitating cultural exchange, in particular, by television broadcasting via satellite (EC, 2003).

Satellite communications might also help to bridge the digital divide in the developing world and between developing and developed countries. Indeed, space-based solutions provide the means to create in short order a fully fledged communication network covering vast territories in countries where terrestrial facilities are underdeveloped or inexistent.

Satellite links are often the only way to bring education to certain regions of the world that are remote and/or lack ground infrastructure. Though distance learning is not a perfect substitute for traditional education, it is a way to help disseminate knowledge and skills to a larger audience. There are still limitations, however, as electronic resources can go only so far in linking learners and teachers. If they are to work effectively, programmes must be well thought out, taking into consideration a country's linguistic diversity (*e.g.* different dialects). For education, this can be achieved through active networking by schools, colleges and universities. Experience in India (presented in the report on the third phase of the project) tends to show that well-managed programmes can be useful in promoting education via satellite in large populations. The main difficulty in developing such applications is the cost of developing the programmes and cost-efficient use of communications satellites links.

Satellite applications can also be used for expanding medical support in developing countries. India represents, again, an interesting case in point. In the 1990s, the Indian government launched a subcontinent-wide telehealth programme to provide medical support to villages via a satellite network, as part of its strategy to develop space applications for sustainable development. The adoption of telehealth via satellite in India has three main drivers. The first is the ability to communicate across the whole country via satellite, thus countering topographic obstacles (*e.g.* terrain, mountains) that hinder the development of land-based communications infrastructure. Another is the possibility for urban-based medical specialists to intervene in remote rural areas, where 620 million Indians live. As an example, there are more neurologists and neurosurgeons in the city of Chennai in South India than in all the states of north-eastern India together. To remedy this "access to medical expertise divide", interactive satellite communications allow specialists to examine patients remotely (via videoconference) and to exchange X-rays and other medical diagnostic data with local doctors. A third is the possibility to eliminate unnecessary travel expenses and strain for patients in poor rural areas. Once the "virtual presence" of the specialist is acknowledged, a patient can access resources in a tertiary referral centre, thereby overcoming the problem of distance (ISRO, at *www.isro.org*).

The Health Channel in South Africa is another example of the use of satellite communications to provide health services in developing countries. It is to be a satellite broadcast channel that delivers information at no charge to patients and healthcare workers in clinics and hospitals in South Africa. The channel was created through a public-private partnership between the South African Department of Health, Sentech, a provider of broadband communications services, and Mindset Network, a partnership led by Liberty and Standard Bank Foundations.[13] The Mindset Health Channel aims to be in all 4 000 public healthcare sites in South Africa within five years, serving 97 000 nurses and 36 million South Africans. Eventually, the channel has the potential to be extended across all of Africa and create a sustainable, mass-scale public health effort tackling all major health issues.

Satellites can also be used to provide medical services to individuals on the move. For instance, in the absence of precise diagnostic facilities on board, pathologies and medical problems encountered in flight frequently require rerouting aircraft. With a portable satellite telemedicine station on board, a diagnosis can be established in flight, avoiding the need to reroute the aircraft in cases of false alarms. If rerouting is required, in-flight diagnosis allows for more timely and effective treatment of the patient, since it provides the medical team on the ground with more precise advanced understanding of the medical problem.

Assessing the benefits of space

The evidence presented in this chapter makes clear that space applications have already been useful for addressing major societal challenges and have the potential to be even more useful in future, if the necessary system upgrading is carried out in a timely manner. However an important question remains: Does space offer effective solutions, when the benefits achieved are measured against costs and when such solutions are compared to alternative terrestrial approaches? The question is difficult to answer. Part of the problem is the fact that credible cost-benefit analyses are not easy to make in the best of circumstances and even more so in the case of space.

Problems arise for estimating both costs and benefits. First, the costs of space systems are hard to estimate, notably because development costs are often not taken into account or unknown. Moreover, a large share of the cost of space systems is fixed; this means that in the case of multi-purpose satellite systems, it is almost impossible to allocate costs meaningfully among different applications.

On the benefit side, it is very difficult to trace the overall societal benefits to space systems (*e.g.* costs of lives saved or property spared thanks to the timely availability of space data in cases of disaster, societal gains in remote areas that are attributable *only* to satellite communications). Space assets – although their role is essential at times – often represent only a small component of the entire socio-economic value chain of a product or a service.

Another difficulty from a decision-making perspective is the fact that technological advances can affect both costs and benefits. Hence cost/benefit studies are often quickly out of date, as technology evolves and the cost-efficiency and capability of systems improve over time. For example, there have been rapid advances in satellite communications and Earth observation technologies in recent years. This means that more can be done (*i.e.* higher benefits) for less (*i.e.* lower costs) with the most recent generation of satellites and that future generations will be even more effective.

Even when credible positive cost-benefit ratios are obtained, one may still wonder whether other uses of the resources spent on the development and operation of space assets would not produce higher payoffs. However, it is not always easy to compare space and terrestrial solutions, and it may be impossible when space offers a unique capability that cannot be duplicated by other means. Moreover, space solutions can rarely be considered in isolation. In most instances, they need to be closely combined with terrestrial facilities to be fully effective (*e.g.* data generated by space-based instruments may need to be integrated in ground-based geospatial models, with complementary non-space data, to be really useful).

These difficulties help to explain why there have been relatively few cost-benefit analyses on space. Those that have been made throw, by and large, a positive light on the question, which tends to confirm more qualitative assessments. This is illustrated here by the examples of El Niño (Box 2.2) and Cospas-Sarsat given above.

Weather satellites represent another of the few cases where the cost effectiveness of space solution can be clearly demonstrated. It is already well known that better weather forecasts resulting from the use of weather satellites generate significant benefits because they enable individuals, public administrations and businesses to cope better with natural hazards such as hurricanes and thus reduce loss of lives and property. Better weather forecasts can also generate significant benefits for industry. Energy utilities for example typically use weather forecasts to determine the energy mix to be used to meet consumer needs. According to the Tennessee Valley Authority in the United States, annual electricity costs would decline by at least USD 1 billion if the accuracy of weather forecasts was improved by one degree Fahrenheit (GEO, 2004).

Cost-benefit analyses are sometimes conducted *ex ante* as a guide for decision making, as in the case of Galileo, the European global navigation satellite system to be deployed in the coming years. The benefits of the system are estimated as the present value of the stream of future revenues that are expected to be generated from commercial applications (*e.g.* location-based services) as well as the public benefits derived from the expected reduction of external costs in air and road transport such as accidents, congestion, air pollution and noise (*e.g.* in-car navigation systems using satellite signals should facilitate traffic management by helping monitor and improve traffic flow).

Paradoxically, the benefits to be derived from space assets may sometimes become clearer when they fail to function properly, forcing users to find alternatives (Box 2.6).

Hence, the cost-benefit studies conducted so far – although partial – tend to confirm the positive overall view of the potential contribution that space might make to society presented in this chapter. It remains to be determined whether this potential has a good chance to be realised, given the conditions under which space activities are carried out. This question will be addressed in the following chapters.

Box 2.6. **The impact of Landsat 7 hardware failure on post fire assessment**

One clue to the benefits of space imaging in fire management can be obtained by examining what happens when a satellite malfunctions. The Landsat 7 hardware failure in May 2003 is an interesting case in point.

Landsat 7 has been used extensively by US forestry officials to identify burned areas, notably in California. When both Landsat 7 and the ageing Landsat 5 satellite were operating correctly, images of fire-ravaged areas could be obtained within eight days, barring cloud cover. Analysts at the US Forest Service's Remote Sensing Applications Center in Salt Lake City would then identify the most severely burned areas by comparing the new Landsat images with pre-fire images. The digital maps produced were sent to soil specialists and hydrologists assigned to burned area emergency response teams. The teams rely on the maps to decide where to lay straw or build erosion barriers to prevent the fires from being followed by another of California's recurring natural disasters, mud slides.

The failure in May 2003 of Landsat 7's Scan Line Corrector, which compensates for the forward motion of the satellite, caused the spacecraft to return images with long data gaps, giving the pictures a venetian-blind-like appearance. With only Landsat 5 providing useful images, the time required to obtain useful images doubled to 16 days. That is too long to help the emergency response teams that must inspect the most severely burned areas within seven days to devise a plan to rehabilitate them.

A possible way to replace the Landsat 7 images appears to be the Advanced Spaceborne Thermal Emissions and Reflection Radiometer (ASTER) aboard NASA's Terra satellite. The Terra satellite passes within about 20 minutes of Landsat 7 on the same orbital track and could ensure coverage every eight days in tandem with Landsat 5. However, while ASTER offers similar infrared data, it does not monitor terrain as routinely as Landsat satellites and it has a narrower swath width. Replacing the Landsat 7 images with scenes from commercial satellites, from US firms or the French SPOT system, forces scientists to bear more costs when acquiring data and causes some unforeseen difficulties. For example, the commercial satellites must be commissioned to return images of specific areas of the globe. That means pre-fire images of particular sites could be hard to obtain.

Source: Iannotta (2003).

Notes

1. Many other countries have decades-old Earth observation capabilities (*e.g.* Canada, Russia, India), and many recent or emerging space countries are in the process of developing their own systems (*e.g.* China, Korea, Turkey).

2. The Norwegian Computing Centre has worked with Statkraft, a major Scandinavian hydropower company, to prepare a snow reservoir mapping system called SnowStar. The SnowStar server automatically processes satellite information of several different systems, including radar imagery, for display within geographic information system software.

3. See, for instance, OECD (2004c), Chapter 4.

4. Farmstar is an example of the application of space technology to precision agriculture. Space images of fields captured by SPOT provide the basis for assessing treatment needs. Such information can then be used for the automatic spraying of the nutrient. SPOT then monitors the results of the operation. Farmstar is currently used for wheat and colza crops. French farms covering altogether 100 000 hectares subscribed to the service in 2004. In the case of colza, the net gain of precision nitrogen spraying alone has been estimated to range between EUR 35 and EUR 75 per hectare.

5. The companies are: BP, DaimlerChrysler, Ford, GM, Honda, Michelin, Nissan, Norsk Hydro, Renault, Shell, Toyota and Volkswagen.

6. In light of current developments, several global independent satellite navigation systems could be operational in the next decade, such as the United States' Global Positioning System (GPS), the Russian GLONASS system, and the European Galileo system. Some regional SBAS are also under development or operational, such as the US WAAS, Europe's EGNOS, India's GPS and GEO Augmented Navigation programme (GAGAN), Japan's Multifunctional Transport Satellite-based Augmentation System (MSAS), and the Canadian Wide Area Augmentation System (CWAAS).

7. The transmission of video, audio and data messages to handheld devices could become an important market for satellites in years to come, especially in remote areas.

8. The role that GNSS can play is recognised in OECD (2004b), which notes that satellite tracking and automatic vehicle recognition systems have the potential to make further improvement to transport charging systems.

9. GPS will provide safety-of-life service only on the Block IIF systems, which have yet to be deployed; hence GPS will not be interoperable with Galileo for this application until full deployment of GPS III.

10. See in particular OECD, 2003a, 2004d and 2004e.

11. DBS may refer either to the communications satellites that deliver the services or to the actual satellite television services. DBS uses special high-powered Ku-band satellites that send digitally compressed television and audio signals to 45 cm to 60 cm (18- to 24-inch) fixed satellite dishes. For the end users of the signals, a DBS reception equipment most often takes the form of a television set-top signal descrambling box, to assure satellite television providers that only authorised, paying subscribers have access to the content. DBS systems transmit signals to Earth in what is called the Broadcast Satellite Service (BSS) portion of the Ku-band between 12.2 and 12.7 GHz.

12. Cable companies have responded by introducing digital cable, which offers more channels than analog cables, and many of the features provided by DBS.

13. Mindset Network is described as a non-profit organisation aimed at the personal, social and economic uplift of all South Africans through better education. It is a far-reaching project that creates, sources and delivers free educational material via satellite broadcasts with supporting multimedia material in print and on the Internet.

Bibliography

Barrett, S. (2003), *Environment and Statecraft: The Strategy of Environmental Treaty-Making*, Oxford University Press, Oxford.

Brachet, G. (2004), "From Initial Ideas to a European Plan: GMES as an Exemplar of European Space Strategy", *Space Policy*, Vol. 20, pp. 7-15.

David, L. (2003), "Bright Future for Solar Power Satellites", *Space News*, 17 October.

Earl, R., G. Thomas and B.S. Blackmore (2000), "The Potential Role of GIS in Autonomous Field Operations", *Computers and Electronics in Agriculture*, Elsevier Science, Vol. 25, No. 1.

EC – European Commission (2001), *White Paper: European Transport Policy for 2010: Time to Decide*, COM(2001) 370 final, Brussels.

EC (2003), *White Paper – Space: A New European Frontier for an Expanding Union: An Action Plan for Implementing the European Space Policy*, European Commission, Brussels.

FAA – Federal Aviation Administration (2004), Satellite Navigation Products Team, http://gps.faa.gov/.

GEO – Group on Earth Observations (2004), "Strategies for Stewardship – Development of a Global Observation System", World Meteorological Organization, 56th Executive Council, 15 June.

Hettling, Jana K. (2003), "The Use of Remote Sensing Satellites for Verification in International Law", *Space Policy*, Vol. 19, pp. 33-39.

ICAO – International Civil Aviation Organization (2001), "Outlook for Air Transport to the Year 2010", Circular 281, June.

IEA – International Energy Agency (2002), *World Energy Outlook: 2002*, OECD, Paris.

Iannotta, B. (2003), "Fire Response Officials Feel Impact of Landsat 7 Glitch", *Space News*, November.

Macauley, Molly K. and Timothy Brennan (1995), "Remote Sensing Satellites and Privacy: A Framework for Policy Assessment", *Law, Computers, and Artificial Intelligence*, Vol. 4, No. 3.

Macauley, Molly K. and Timothy J. Brennan (2001), "Private Eyes in the Sky: Implications of Remote Sensing Technology for Enforcing Environmental Regulation", pp. 310-334, in Paul S. Fischbeck and R. Scott Farrow (eds.), *Improving Regulation: Cases in Environment, Health, and Safety*, Resources for the Future, Washington, DC.

Mistra, A. *et al.* (2004), "Navigation for Precision Approach", *GPS World*, 1 April.

Nijkamp, P. *et al.* (1998), *Transport Planning and the Future*, John Wiley and Sons Ltd., London.

NOAA – National Oceanic and Atmospheric Administration (2004), *Economic Statistics for NOAA*, Third Edition, Washington, DC.

NRC – National Research Council (2001), *Laying the Foundation for Space Solar Power – An Assessment of NASA's Space Solar Power Investment Strategy*, Report of the National Academies, Washington, DC.

OECD (2001), "Environmental Strategy for the First Decade of the 21st Century", adopted by OECD Environment Ministers, 16 May, OECD, Paris.

OECD (2003a), *Emerging Risks in the 21st Century: An Agenda for Action*, OECD, Paris.

OECD (2003b), *The Sources of Economic Growth in OECD Countries*, OECD, Paris.

OECD (2004a), "Outcomes of the Meeting of the Environment Policy Committee at Ministerial Level", OECD, Paris, 20-21 April.

OECD (2004b), "OECD Environmental Strategy: 2004 Review of Progress, Policy Brief", *OECD Observer*, OECD, Paris, April.

OECD (2004c), *Space 2030: Exploring the Future of Space Applications*, OECD, Paris.

OECD (2004d), *Large-scale Disasters: Lessons Learned*, OECD, Paris.

OECD (2004e), *The Security Economy*, OECD, Paris

UN – United Nations (2002), "Report of the World Summit on Sustainable Development, Johannesburg, South Africa, 26 August-4 September 2002", A/CONF.199/20, New York.

UNFCC – United Nations Framework Convention on Climate Change (2004), Kyoto Protocol, *http://unfccc.int/2860.php*.

WBCSD – World Business Council for Sustainable Development (2004), "The Sustainable Mobility Project, Mobility 2030: Meeting the Challenges to Sustainability", Overview 2004, July.

Zhang, Y. et al. (2002), "Application of an Empirical Neural Network to Surface Water Quality Estimation in the Gulf of Finland Using Combined Optical Data and Microwave Data", *Remote Sensing of the Environment* 81, pp. 327-336.

Chapter 3

Supply Conditions: Strengths and Weaknesses of Space

> *While space holds a great deal of promise for society at large, it is far from clear whether this potential will actually be realised, given the state of the space sector and the major technological and economic challenges space actors will have to overcome in coming decades.*
>
> *Today the space sector offers a mixed picture. The industry's upstream segment (manufacturers of launchers and spacecrafts) still has not recovered from the downturn of the early 2000s. The downstream segment is in a stronger position, although some markets have not developed as expected (e.g. Earth observation) while others are only emerging (e.g. navigation).*
>
> *The sector's future is thus far from certain. Although space has a number of inherent strengths, it also faces important technological hurdles, notably regarding access to space. This is reflected in high costs and long lead times for the development of new systems. Space solutions also have to contend with strong competition from terrestrial technologies in some applications (e.g. broadband, navigation, Earth observation), while the dual civil/military nature of space technology is a mixed blessing.*
>
> *Whether these challenges can be overcome will very much depend on whether the framework conditions (i.e. institutional arrangements as well as laws and regulations) that govern space activities provide a policy environment that is sufficiently supportive for the development of space applications that effectively meet users' needs in a predictable and sustainable manner.*

3. SUPPLY CONDITIONS: STRENGTHS AND WEAKNESSES OF SPACE

Introduction

The two previous chapters have shown that space applications hold a great deal of promise, but it is not clear that their potential will in fact be realised.

First, as the initial phase of the project made clear, the space sector currently offers a mixed picture: its upstream component is subject to chronic excess supply conditions; while its downstream component is unevenly developed. Moreover, space business is not business as usual. In contrast with other sectors of the economy, governments continue to play a dominant role; they conduct the basic research and development (R&D) and develop new systems; acquire space goods and services from the private sector; and set the framework conditions that govern space activities. Moreover, because of the dual civil/military nature of space technology, governments pay particular attention to space firms and impose security-motivated restrictions (*e.g.* export controls) on their activities.

Second, further development of space applications faces serious technological and economic challenges. Space technology does have inherent strengths that make it particularly suitable for developing systems to help governments meet future challenges. However, this will require continuing technological advances, not an easy task given the economic constraints faced by space actors. A key condition of success – independently of the question of resources – will be the existence of an institutional, legal and regulatory framework that fully supports the development of the needed space systems and their effective use.

This chapter first briefly reviews the current state of the space sector, as revealed in the first phase of the project. It will then consider, from a technological and economic perspective, the strengths and weaknesses of space for responding effectively to future challenges. Finally, on the basis of work conducted by the OECD in recent decades, it reviews the key role played by framework conditions for economic and social development in general, providing a useful backdrop for considering the main findings of the third phase of the project regarding factors that are critical for the success of space applications. These findings clearly confirm the importance of framework conditions for space and provide a concrete basis for the detailed discussion of this issue in subsequent chapters.

The current state of the industry

After a phase of rapid expansion in the 1990s, the space industry experienced in the early 2000s a severe downturn in the aftermath of the dot.com bubble and of "the collapse of the big LEO", *i.e.* the failure of the large constellations of low Earth orbit (LEO) mobile telecommunication satellites, such as Iridium, to meet the expectations of their sponsors. The upstream component (launcher and satellite manufacturers and providers of launching services) was particularly hard hit. Downstream firms (providers of space-based products and services, notably telecommunications, positioning and navigation services and Earth observation services) have fared better.

Slow recovery upstream

Launching activities remained rather depressed in 2004 for the fourth year in a row. Currently, many actors in the launch industry are facing a difficult situation, in which the number of launchers exceeds the satellites earmarked for launch. Aside from the current relatively flat commercial market, the situation is exacerbated by significant progress in terms of the durability and capacity of spacecraft, which has reduced the need for new satellites and their replenishment. In 2004, there were 63 launches worldwide, compared with the 62 performed in both 2002 and 2003 (Edwards, 2005). Out of those 63 launches, fewer than 20 were commercial ones, showing the importance of governmental markets for both satellite manufacturers and launch providers.

Like the launcher industry, the satellite manufacturing industry suffered in the early 2000s. In 2001 only 75 satellites were launched, the lowest number in the past decade and a 32% drop from the previous year. While just over 80 satellites were launched in 2002, the number dropped back to 69 in 2003 (for purposes of comparison, 150 satellites were launched in 1998). According to Euroconsult, only 19 commercial satellites were ordered in 2003 at a total estimated value of USD 2.1 billion.

In future, the commercial satellite market may pick up somewhat as a result of the deployment of high-definition television (HDTV) and the replenishment needs of several satellite fleets. However, services such as broadband and Internet via satellite have been slower to emerge than expected. Moreover, consolidation among satellite operators that face strong competition from cable television and fibre networks helps to reduce the demand for new spacecraft. Over the next ten years, about 236 commercial communications satellites destined for geostationary or medium Earth orbit may be built at an estimated value of USD 26 billion, according to Forecast International. Over the same period, the low Earth orbit market, essentially for mobile communications, will see production of 34 spacecraft worth USD 154 million (Edwards, 2005).

3. SUPPLY CONDITIONS: STRENGTHS AND WEAKNESSES OF SPACE

Faced with sluggish growth on the commercial front, the launching and satellite manufacturing industry should continue to rely heavily on government contracts. The military market is expected to be a major contributor to the recovery. Indeed, military contracts offer lucrative, long-term work to contractors who are seeing their commercial business dry up. This will benefit most directly the major US providers of military space equipment, as the United States accounts for 95% of the world's military space expenditures. Indeed, in the United States alone, military contracts scheduled for the next ten years but not yet awarded could be worth more than USD 15 billion. Particularly lucrative for US manufacturers will be the massive Global Positioning System (GPS) and Transformation Communication programmes. The situation is much bleaker for manufacturers in Europe and Japan where governments continue to cut costs – and their satellite production requirements – by pooling space resources. Sharing data or satellite tasking time allows European countries to multiply their satellite capabilities without added expense. Major European and Asian manufacturers should nevertheless manage to remain afloat as they await the expected rebound of the commercial market (Edwards, 2005).

Uneven growth downstream

The situation looks somewhat brighter downstream, although growth is uneven. Revenues of the 36 communication satellite operators that make up the fixed communications satellite services industry, the most mature downstream component, remained flat in 2003 at USD 6.15 billion. These services represent 95% of total satellite communications revenues.[1] While communications satellite operators have not yet fully recovered from the dot.com bubble, they have benefited from rapid progress by their main clients, the providers of direct broadcasting services (DBS), which account for two-thirds of their revenue.[2] In fact, the world DBS industry has exploded, rising from USD 1.5 billion in 1995 to USD 22.5 billion in 2001, when more than 54 direct-to-home (DTH) platforms distributed more than 5 000 TV channels to over 45 million subscribers around the world. In 2003, revenues of the 54 companies that make up the industry rose to USD 33 billion, an increase of 27% over the previous year. The direct broadcasting of radio by satellite to moving vehicles is also proving successful in the North American market, although on a more modest scale (4.3 million subscribers by the end of 2004). Satellite broadband is an emerging application that may provide an effective solution to meeting the needs of users in rural and remote areas in the coming years. However, this remains a niche market and progress has been slow.

Another market segment that is experiencing rapid growth is satellite-based location and navigation services, although only one such system, the US GPS, is fully operational globally today.[3] The entry of the European system Galileo in the second half of the decade should further spur market growth.[4]

Optimists even predict that by 2020, 2.5 billion people will be using navigation systems. In 2003, the global product revenues for navigation services were estimated at EUR 15 billion, a figure expected to rise to EUR 47 billion by 2005 and EUR 178 billion by 2020, as navigation chips are integrated in more and more products (GMO, 2003).[5]

Earth observation (EO), the third main component of the downstream segment, is much smaller and struggling. While EO is one of the oldest satellite applications, commercial observation satellites (COS) are still relatively new. The industry only started up when restrictions on satellite imagery technologies were relaxed at the end of the cold war. Despite substantial technical progress in recent years, the economic prospects of COS remain uncertain in a very competitive market. In 2003 sales by the commercial remote sensing industry, including aerial and satellite segments, were estimated at USD 2.6 billion, with the satellite segment representing roughly a third of the total. By 2010 sales may reach USD 6 billion with USD 2 billion for the satellite segment.

Key role of public space markets

While commercial demand for space products and services has grown in importance over the years, governments still represent a major market for the space industry. Indeed, following the downturn in commercial activities since 2000, they have regained their leading position. In 2001, world public budgets for space activities were estimated at about USD 38 billion; they rose to USD 43 billion in 2003 and may exceed USD 50 billion by 2010. In 2003, about 57% of public space resources were devoted to civil applications (USD 24.3 billion), with the remaining 43% (USD 18.5 billion) allocated to military space programmes. By the end of the decade, military space budgets may reach a level similar to those of civil programmes for the first time since the end of the cold war (Euroconsult, 2004).

Particularly significant for the future of the space sector is the expected growth and reorientation of US public space budgets. The US military space budget is expected to rise from USD 17.5 billion in 2003 to an estimated USD 25 billion in 2010, a 40% increase. Under President Bush's new space exploration plan, announced on 14 January 2004, NASA's budget may also grow, although more slowly (possibly by 5% a year for the next five years), and may reach USD 18 billion by 2010 (USD 15.9 billion in 2005). At the same time, funds are expected to be substantially reallocated from the space shuttle (to be retired by 2010) and the International Space Station (ISS) (to be completed in 2016) to exploration missions and the development of a crew exploration vehicle.

The more modest European consolidated space budget (around EUR 5 billion or USD 6.2 billion in 2003) should also expand, but at a slower pace, and could reach some USD 8 billion by 2010. In 2005, the budget of the European Space

Agency (ESA) represents EUR 2.9 billion (approximately USD 3.8 billion). Several European countries have their own national programmes, although more often than not, large parts of the funding are destined for ESA programmes. For instance, the budget of the French Centre National d'Études Spatiales (CNES) is EUR 1.3 billion (USD 1.5 billion) in 2005, out of which around EUR 685 million (USD 908 million) goes to European programmes.

In Asia, space prospects vary from country to country. In Japan, the space budget has declined over the past few years, and space activities still need to recover from several launch and satellite failures. However, more resources may be devoted to space in the coming years, in response to recent geopolitical concerns in the region (JAXA's budget amounts to USD 2.48 billion in 2005). Rapid growth is expected in the public space budget of several Asian space-faring countries, notably China and India, although from a much lower base. For its part, Korea is investing not only in the development of its own space launch infrastructure, but also in human spaceflight by purchasing a seat for its first astronaut on the Soyuz flight to the International Space Station (to be launched no earlier than 2007). Finally, new emerging countries are developing their own space capabilities (*e.g.* Turkey, Malaysia, Iran, Pakistan, Nigeria).

The levels of institutional space activities differ widely among countries/regions, as do the specific situations for the growth of an indigenous space industry. However, when comparing available institutional funding, actual purchasing power parity and local labour costs need to be taken into account (for example, it is generally less expensive to have Chinese or Russian engineers work on a project than Americans or Europeans).

Hence, on balance, the space sector offers a mixed picture today. The upstream segment still generally suffers from weak market conditions, as the commercial segment of the market has not yet recovered from the dot.com bubble. This has forced space actors to rely more heavily on public markets. Things are somewhat different in the downstream segment. There is rapid growth in some segments (satellite communications and navigation), but performance remain disappointing in the Earth observation segment. The greater reliance expected in future on the public market (both military and non-military) means that, over the coming years, the sector's fate will be largely determined by public action. In this regard, one may wonder whether the greater emphasis on military space and space exploration, although these are clear drivers for technological progress, will not be at the expense of socio-economic applications.

Space technologies: strengths and challenges

Space technology has been the object of extensive studies over the years. The following sections build upon those findings and on the work carried out in the course of the project, notably in the third phase, to provide a general summary of what appear to be the major technical characteristics of space technologies from an overall economic perspective.

Space systems' inherent strengths

The development of space systems takes advantage of the inherent strengths of space technology, namely:

- *Inherent wide-area observation capability* offering synoptic views of large-scale phenomena and placing *in situ* measurement in the global context required, among others, for the observation of many environmental and climate phenomena.
- *Non-intrusive observation* allowing collection of data without compromising national sovereignty as ground-based measurement or air-borne remote sensing might. This is an advantage, for instance, for the monitoring of international environmental treaties. Moreover, data can be collected via satellite over sites which cannot be accessed by other means, *i.e.* sites that make *in situ* measurements too difficult.
- *Uniformity* in that the same sensor may be used for many different places in the world, thus helping to ensure that the data collected in this way are comparable, as they are generated by the same instrument.
- *Rapid measurement capacity* allowing sensors to be targeted in relatively short order at any point on Earth, including remote and hostile areas.
- *Continuity* with single sensors or series of sensors providing long time series that can be collected over the lifetime of the spacecraft. Such continuity is particularly important, for instance, for climate studies.
- *Dissemination of information over broad areas* as communications satellites have the capability to broadcast hundreds of television channels over continents.
- *Rapid deployment* as communications satellite can be rapidly deployed over areas where terrestrial networks have been impaired or are insufficient to meet information needs.
- *Global navigation* since global navigation satellite systems (GNSS) offer a unique capability to establish the location of fixed and moving objects anywhere in the world, complementing the information provided by EO satellites, for instance, for the development of geographic information systems (GIS), a key foundation for planning and decision-support systems.

3. SUPPLY CONDITIONS: STRENGTHS AND WEAKNESSES OF SPACE

Main technological challenges

However, it is not easy to take advantage of these strengths. The space environment is hostile for equipment and even more so for humans. This means that space missions require the design, construction and thorough testing of complex equipment, capable of withstanding extreme conditions in space and with sufficient built-in redundancy to minimise the chances of failure. This translates into long lead times and large expenditures for the deployment of space systems.

Access to space

Sending payloads into space is a formidable challenge. Major progress has been achieved over the last few decades, including notably the successful development of several families of launchers (*e.g.* Soyuz, Ariane, Atlas, Delta). However, the technology has not developed as originally hoped: access to space remains costly and risky. Efforts to move from expendable to reusable launchers have not been successful.[6]

The slow progress of space technology can be explained in part by the high risks involved, not only for space actors, but also for the population at large, as well as by strategic considerations. Because of such risk and strategic factors, new space-related technical developments are subjected to extensive public scrutiny. Although there are good reasons to put in place such controls and regulations, they represent a major cost and lengthen the time required to get a payload into space (including the development of new launchers), notably for new entrants. Strategic considerations are another major factor. The main objective of space programmes is to provide ensured access to space; however, transfers of technologies are restricted, so that the objective is best achieved by developing a launcher that uses mature technology. Hence, the only technology that can be used effectively today (*i.e.* expendable launch vehicles, ELVs) has evolved little over the last half century. Indeed, in 2004, the venerable Soyuz was still the world's most actively used launcher, and, until the shuttle is back in service, it is the only human-rated vehicle capable of servicing the International Space Station.

According to many experts, a paradigm shift and perhaps some changes in existing rules of the game are needed to achieve significant progress in reducing costs. Conceptually, the best way to achieve a substantial reduction in the cost of access to space in the longer run is to develop reusable launch vehicles (RLVs).[7] However this requires not only major progress in space technologies (*e.g.* space propulsion, space transport systems and orbital systems) but also advances in enabling technologies (*e.g.* electronics, communications, nanotechnology, artificial intelligence, new materials, biotechnology).

One can envisage, for instance, that progress in propulsion will gradually lead to rocket engines that are more cost-efficient, reliable and robust, making progressively possible the recovery and reuse of engines or whole transport systems.[8] Following the space shuttle experience and some ambitious programmes in the 1990s (*e.g.* X–33, European Hermes), it became clear that new technical developments would take time to emerge, while necessitating continuing political will to sustain a reasonable level of R&D. A hopeful sign in this regard is the progress recently achieved in the development of air-breathing engines such as scramjets (*e.g.* NASA's X-43A demonstrator which flew at Mach 7 in March 2004 and close to Mach 10 in November 2004, beating the speed record for an air-breathing engine), while military R&D, in particular in the United States, may also provide some new technical avenues in the next decade.[9]

While such R&D efforts continue to prepare for the future, the market is likely to be dominated by the current generation of expendable launchers for the large majority of missions. Although they are expensive, these launchers are by definition well-suited for one-way missions (*i.e.* launching a satellite into space), which represent by far the largest segment of the launch market. It follows that the main development efforts over the next few decades may be a challenging two-track effort involving long-term R&D work to develop RLVs and continuous efforts to bring down the cost of ELVs.

Telecommunications

It is generally agreed that the age of global satellite communication started in 1965 with the launching of COMSAT's first geostationary satellite Early Bird, which provided both telephony and television services. Over the years, major technological advances have been achieved, resulting in drastic declines in cost and huge increases in capacity.[10]

In recent years it is perhaps in broadcasting that the most significant progress has been achieved, including notably the development of direct-to-home (DTH) platforms, which make it possible to deliver digital video and audio services via satellite directly to the home of users. Other segments of the market have experienced setbacks as a result of rapid progress in competing terrestrial technologies (*e.g.* fibre optics, mobile phones). As a result, communications satellites compete very effectively with terrestrial broadcasting services, but they are often more costly than terrestrial alternatives for two-way communications in most markets, and latency reduces their usefulness for services requiring instant interactivity. Moreover, the current and future Ka-band systems, which aim to provide high-speed, two-way data communications for Internet and other services, need to overcome the challenge of ensuring transmission of the high-frequency signals through rain.[11]

Future major satellite communications efforts are likely to focus on the development of applications such as high-definition television (HDTV) and interactive television (iTV), where satellites have a comparative advantage and/or can take advantage of complementary technologies.[12]

New technology (*e.g.* spot beams, onboard processing) currently being deployed (*e.g.* Alenia Spazio's SkyPlexNet, WildBlue) or under development (*e.g.* the Japanese WINDS project described in Box 3.1) should overcome some of these weaknesses. They should allow DTH services operators to offer "double play" (television channels plus broadband) on the same dish, and Internet providers to offer broadband in niche markets. These include, for instance, services designed to meet the communications needs of highly mobile people or to provide communications support in emergency situations. Many operators are looking to achieve technical breakthroughs in this domain.[13] Based on these and other recent technological developments, many satellite operators have started to set up partnerships with manufacturers and Internet providers to provide two-way satellite broadband service solutions to enterprises, small office/home office users and local communities (*e.g.* SES Global's partnership with Gilat and Alcatel Space in the Satlynx joint venture).

In developing countries, satellite communications can allow the rapid deployment of communications services when terrestrial alternatives are underdeveloped, thereby contributing to economic and social development, for example by providing distance education and telehealth services.

Box 3.1. Japan's R&D efforts in satellite broadband

In Japan, important space R&D projects have been launched as part of the government's e-Japan Priority Policy Programme, whose goal is to make Japan the world's most advanced nation in information technology by 2005. The Wideband Internetworking Engineering Test and Demonstration Satellite (WINDS) is a joint development of JAXA and the National Institute of Information and Communications Technology. Under the overall strategy, the demonstration satellite to be launched in 2005 and its telecommunication system (WINDS) are to provide ultra-high-speed Internet (up to 1.2 Gbps) domestically. In addition, the project aims to provide ultra-high-speed international Internet access in several Asia-Pacific countries and regions, and there is active collaboration with several Asian universities and technical centres. The satellite will carry two Ka-band multi-beam antennas delivering fixed satellite communications throughout Japan and major Asian cities (Hong Kong, China; Kuala Lumpur; Singapore; Manila; Jakarta; Bangkok; Bangalore; Seoul; Beijing; and Shanghai).

Source: JAXA (2004).

In the longer run, satellite communications may be displaced by optical fibre in some markets, as terrestrial infrastructure gradually expands. However, the necessary investments are enormous and it is not clear if and when they are likely to be profitable.[14]

Earth observation

Major progress in Earth observation has been achieved over the years. The development of EO systems was largely driven originally by geopolitical considerations. However, as discussed in Chapter 2, the range of applications has increased considerably over time to include weather monitoring and prediction, environmental research and monitoring, the management of natural resources, agriculture (*e.g.* agriculture policies in the European Union), urban planning and environmental protection, and of course, disaster preparedness and mitigation.

Despite these impressive achievements, there are still gaps in existing technical capabilities that reduce the usefulness of EO data, notably in terms of resolution and revisit frequency. Other increasing problems include the processing of large amounts of data, their integration into useful information and archiving.

One of the challenges is to meet to the daily needs of users, not only scientific organisations but various governmental agencies, international organisations, local planners and private users (*e.g.* fishermen, farmers) (ICSU, 2004). For instance, despite the huge increase in the amount of climate and environmental data delivered by NASA's Earth Observing System (EOS) (from 17 terabytes in 1999 to 1 000 terabytes in 2004), the scientific community is divided over the actual cost-worthiness of the system, as it is feared that the data may not be fully exploited.

Efficient information systems and communications links that provide comprehensive data to the user community are essential for the effective utilisation of space instruments, but this has been often underestimated. Other challenging aspects are the accuracy and reliability of the data, as satellite signals tend to vary over time and most missions are still experimental and short-term (*e.g.* the Envisat mission should last for only a little over five years), while scientists often need long time series of consistent data, notably for environmental research (Lawler, 2004).

These technical challenges are being increasingly addressed. Current R&D on constellations of small satellites and their instruments may provide platforms that may become as efficient as today's larger, more expensive platforms. Instrument-wise, the development of new synthetic aperture radar (SAR) systems using X-bands, future interferometric SAR (InSAR) systems, superspectral and hyperspectral systems will also provide interesting new data for many sectors.[15] Finally, in the coming years, the development of revolutionary high-resolution geostationary satellites may make possible continuous and detailed observation of large areas of the Earth (*e.g.* areas prone to natural disasters).

The distribution of space-based data in integrated information systems is also a key technical challenge that is gradually being addressed. Lower engineering costs are making direct broadcasting more easily achievable. This, in turn, should enhance the opportunities for developing EO applications. However, some of the limitations of existing systems (for applications like emergency services) will be hard, if not impossible, to overcome (*e.g.* real-time reactivity). Moreover, Earth observation may have to face increasing competition from aerial observation technologies. Indeed, the increasing digitisation of aerial data and the technical improvement of those systems (*e.g.* better drones) may make them better than satellites for mapping small areas.

Finally, major efforts will be needed to put the production and distribution of space-based EO data on a sustained operational footing so as to ensure that the data are produced on an ongoing basis and fully meet users' requirements, notably for applications (*e.g.* environment, management of natural resources, security) that are important for the future.

Global navigation satellite systems (GNSS)

Navigation systems, like many other space applications, have a military origin. For instance, the US Global Positioning System (GPS), the main system in operation today, was developed by the US Department of Defense (DoD) and deployed over two decades at an estimated cost of over USD 10 billion. GPS is a major technical achievement. Beyond its military use, it has proved its effectiveness in a number of civil and commercial applications. Indeed, GPS is today a worldwide information resource that supports a wide range of civil, scientific and commercial functions, from air traffic control to the Internet. It has spawned a substantial commercial industry in the United States and abroad, with rapidly growing markets for related products and services.

However satellite navigation systems are not yet completely suitable for applications requiring high levels of accuracy and reliability. Satellites transmit relatively weak, high-frequency signals that are subject to interference from disturbances in the ionosphere or from radio frequencies. The signals can be jammed, either intentionally or otherwise (US DOT, 2001). They also have some inherent limitations is terms of technical accuracy, as signal failure may occur in cities, forests, mountainous areas, tunnels or underground car parks. Under certain conditions (*e.g.* in urban areas, inside buildings), their use is limited and there are concerns that the integrity of the signal may not yet be sufficient for applications where this is vital (*e.g.* air traffic control).

Many applications of satellite navigation systems are still experimental. This is the case, for instance, of the road traffic management applications that were considered in the third phase of the project. While this type of application has been successful in Switzerland for controlling transit traffic of trucks, a German attempt to introduce a road-charging system using satellite technologies was abandoned at the end of 2003 (because of political and technical difficulties), after investments of around EUR 700 million. In the United Kingdom, the Congestion Charging Scheme in central London is not GNSS-based but has used automatic reading of automobile number plates since 2003. However, in 2006, satellite technologies may be used nationwide to charge heavy goods vehicles. This may be a first step towards more widespread use of satellite technology to reduce road transport congestion.

Progress is under way to overcome the current weaknesses of these systems. The major technological challenges are to improve the reliability, accuracy and integrity of positioning systems. In this regard, the development of new differential GPS (DGPS) systems, such as the US Wide Area Augmentation System (WAAS) and the European Geostationary Navigation Overlay Service (EGNOS), and new generations of interoperable satellite constellations (*e.g.* Galileo and GPS III) should significantly increase the quality and robustness of the signal available in most parts of the world. However, as was found in the third phase of the project, space-based navigation systems may face increasing competition from terrestrial alternatives in the coming years (*e.g.* mobile telephony networks, enhanced cell ID-based technologies). The two technologies could also be combined in some applications (*e.g.* urban transport) to provide a more robust navigation system.

Human presence in space

Establishing a human presence in space has been an important objective of the space programmes of major space-faring countries (*e.g.* race to the Moon, Skylab, Mir, International Space Station, European Astronaut Corps, Chinese spaceflight). This segment of space utilisation will continue to expand in the years to come for both political and scientific reasons. Moreover, some new commercial markets may emerge (*e.g.* space tourism ventures).

Many technical achievements have been made since 1957 and the first flight of Yuri Gagarin, especially in terms of space stations.[16] The past decades have shown that the basic technical requirements for sustaining humans in orbital stations for limited periods of time have been satisfied (*i.e.* some Russian cosmonauts have spent up to a year in orbit).

However, aside from the development of cost-effective human-rated launchers, this high-risk activity faces a number of important challenges. New materials are needed to counter threats from space debris and provide better protection against potentially deadly cosmic radiation, especially to ensure safer flights of long duration. Better environmental control and life support systems are needed to remove airborne contaminants (such as carbon dioxide) efficiently and ensure renewal of oxygen and water. The system currently mostly used, the "open loop system", means that all food, water and oxygen have to be brought aboard the ISS at great expense. Increased research on closed loop systems should eventually reduce the need to resupply orbital outposts. More orbital research may be needed on the effects of radiation to contribute both to the study of patients' medical conditions on Earth and to potential future human exploration (i.e. when leaving the protective shield of Earth's magnetic field for longer and longer periods, cell damage from cosmic radiation is theoretically inevitable).

Current efforts in space agencies and the commercial sector may pave the way for possible innovative and more cost-effective solutions (e.g. inflatable in-orbit structures developed by Bigelow Aerospace in cooperation with NASA, the Inflatable Habitat programme led by the European Space Agency).[17]

Maintenance of space assets

Maintenance operations in space include the servicing of space platforms (e.g. satellite, space station) to replenish consumables and degradables (e.g. propellants, batteries, solar array); replace failed functionality (e.g. payload and bus electronics, mechanical components); and/or enhance the mission (e.g. software and hardware upgrades). They could also logically include the orderly disposal of satellites at the end of their useful lives. Today, maintenance of space assets is very limited, as it is technically challenging and requires, in some cases, very costly human intervention. However, over the long run, significant progress could be made towards developing a viable and cost-effective robotised space assets management system if adequate resources are devoted to this goal.

Several countries have developed or are in the process of acquiring some basic capabilities for in-orbit servicing. First, in terms of autonomous docking in-orbit mechanisms, Russia has relied successfully for years on the ability of its Progress and Soyuz vehicles to dock to different space stations. In 2005, Europe should launch the cargo ship Jules Verne, the first of its automated transfer vehicles (ATVs), to the ISS (Box 3.2). The successful space shuttle servicing missions to repair and enhance the Hubble Space Telescope have provided some useful experience, although they have been very expensive (about USD 500 million each). The robotic Hubble servicing mission under consideration in 2004 would have been even more costly (by some estimates it would have cost more than USD 2 billion).[18]

> Box 3.2. **Autonomous in-orbit rendezvous and docking: first steps for in-orbit servicing capabilities**
>
> To undertake in-orbit servicing of orbital infrastructures will require overcoming important technological hurdles to be able to rendezvous and dock autonomously in orbit.
>
> **ESA's family of cargo ship ATVs.** The European Space Agency's automated transfer vehicle is an expendable cargo ship to be launched in late 2005 atop an Ariane 5 launcher, to carry 7.5 tonnes of dry cargo and fluids to the International Space Station. Once connected to the ISS, the ATV re-boosts the station to a higher altitude to compensate for atmospheric drag and removes waste from the station when it is detached. The ATV, developed by EADS Space Transportation and its subcontractors, uses critical robotics capabilities to perform automatic rendezvous and docking. It is a fully automated spaceship with a multiple-fault-tolerant capability, although the ISS crew can initiate a collision avoidance manoeuvre, in case of malfunctions, to move the spaceship away from the ISS during the rendezvous phase. The 20.7 tonne ATV, which should conduct regular missions to the ISS, has about three times the payload capability of its Russian counterpart, the Progress-M cargo vehicles.
>
> **NASA's DART demonstrator.** NASA's DART (Demonstration for Autonomous Rendezvous Technology) is a flight demonstrator vehicle designed to test technologies required to locate and rendezvous with other spacecraft. It is completely autonomous, and the entire 24-hour mission should be accomplished without human intervention at a cost of USD 95 million. Developed by Orbital Sciences Corporation of Dulles, Virginia, the DART vehicle should be launched in 2005 on a Pegasus rocket to test rendezvous, close proximity operations and its control between the vehicle and a stationary satellite in orbit.
>
> *Source:* European Space Agency (2004c), NASA Marshall Space Flight Centre (2004).

The technological challenges are many. Aside from automated docking capabilities, in-orbit servicing requires the capacity to conduct proximity operations, which involves not only having robots able to perform the required tasks technically, but also to be capable of remaining close enough to the spacecraft to be serviced or repaired to do the work effectively. This is a major challenge in itself. When in orbit, space platforms can move at speeds of several kilometres a minute, depending on their altitude, and it is quite difficult to have several spacecraft "flying" very close to each other. The XSS-11 (experimental small satellite) microsatellite, developed by Lockheed Martin Space Systems and funded by the US Air Force Research Laboratory, is an interesting project to be launched in 2005. Its objective is to demonstrate a microsatellite's extended proximity operations with a spacecraft already in orbit (i.e. rendezvous, standoff inspection and circumnavigation) (Berger, 2003).

With respect to the capacity of robots to perform the required tasks, several options were under investigation for a possible robotic mission for servicing Hubble. Those options, if not actually used for the Hubble telescope, provide some interesting avenues for future in-orbit servicing activities. One idea is to use very dexterous tele-robots assisted by humans on the ground rather than autonomous robots. Possible candidates include the Johnson Space Center's Robonaut and the University of Maryland's Space Systems Laboratory's Ranger robot. Robonaut is a human-like android designed by the Robot Systems Technology Branch at Johnson in a collaborative effort with the Defense Advanced Research Projects Agency (DARPA). The Robonaut project is focused on developing and demonstrating a robotic system that can perform the same duties as a spacewalking astronaut. The University of Maryland's Ranger robot is flight-ready, according to its designers, and has dexterous manipulators capable of working on Hubble. The Ranger robot has already undergone testing against Hubble servicing tasks, according to project personnel (NASA, Johnson Space Center, 2004). In Europe, the Eurobot programme's first phase of development is being carried out for ESA by a consortium led by Alenia Spazio. Eurobot, a robot as large as a human, is designed to carry out astronauts' tasks on the ISS, and may in time be able to climb outside of the station, attach itself to the handrails like an astronaut and be tele-operated by the crew inside (ESA, 2004a).

For NASA, a major perceived advantage of adopting a robotic solution for servicing Hubble, based on the different current demonstration projects, was that the technology needed to carry out space-based robotic repair fits neatly with President Bush's vision of developing robotics and other capabilities necessary for setting up a Moon base and sending astronauts to Mars.[19] A similar approach is followed in Europe in order to develop innovative robotic solutions for future space exploration with the Aurora programme (ESA, 2004b).[20]

Demonstrating the technological feasibility of robotic servicing missions is only the first step. Spacecraft also have to be serviceable at reasonable cost, and an appropriate infrastructure to carry out the servicing has to be available. In this regard, the objective of the Orbital Express Advanced Technology Demonstration Programme, developed by Boeing and Ball Aerospace and funded by DARPA, is to demonstrate autonomous techniques for in-orbit refuelling and reconfiguration of satellites. Two satellites are to be launched in 2006: the service vehicle, ASTRO, and the Next Generation Satellite and Commodities spacecraft (NEXTSat/CSC). The objective of this programme is also to start developing industry standard servicing interfaces and protocols for use by future spacecraft developers (Ball Aerospace, 2004).

While promising technologies are under development, it will be some time before robotic servicing is fully technically feasible, and even longer before such servicing become economically feasible. As a result, almost no planned commercial satellites are currently designed with servicing in mind.

Economic challenges

Space entrepreneurs currently face economic challenges, on both the supply and the demand sides. On the supply side, the challenges are predominantly technology-related. They include notably the high cost of access to space, the long lead time of space projects (due to the complexity of systems and the practical inability to fix systems once in space) and the role of economies of scale in production. While technologies vary from application to application, these challenges are common to all.

On the demand side, the challenges are more application-specific; they relate to the nature of the output produced by the application (*e.g.* whether the space-based product or services are public or private goods) and the conditions under which demand arises.

Another important factor is the central role played by government. On the supply side, governments encourage and contribute to the development of applications, in order to meet a number of public strategic and socio-economic objectives. On the demand side, public agencies are major buyers of space goods and services. At the same time, governments play the principal role in the regulation of markets (*e.g.* maintaining a level playing field for space goods and services).

High cost of access to space

Although significant progress in space technology has been achieved over the years, current space transport systems remain extremely costly and risky. Typically, the cost of placing 1 kg of payload in low Earth orbit is around USD 10 000 today and has declined little over the last few decades. This high cost results from the fact that: i) launch vehicles are expensive to develop and are developed over several years; ii) production runs are short, so that the high development costs are amortised over a small number of vehicles;[21] and iii) the vehicles are expendable, i.e. they are only used once.

For purposes of comparison, the cost of developing a large launcher is of the same order of magnitude as that of developing a large airliner (i.e. USD 5 billion to USD 10 billion). However, while for aircraft, the cost is amortised over hundreds of vehicles, it is amortised over tens of launchers. Therefore, on a per vehicle basis, the development cost per aircraft is roughly one-tenth of what it is for launchers. Moreover, while airliners are designed to perform hundreds of flights a year, expendable vehicles by definition perform only one.

3. SUPPLY CONDITIONS: STRENGTHS AND WEAKNESSES OF SPACE

Launchers require not only large upfront investments but also long-term commitments to fund the launchers in case of failures and when necessary upgrades are needed.[22] This translates into a poor financial return on investment:

- In Europe, the Ariane 5 launcher's cumulative development costs already represent more than USD 7.5 billion, but the programme had to bear increased costs associated with the failure of Ariane 5's first flight in June 1996 and of the new Ariane-5-ECA in December 2002.

- In the United States, the evolved expendable launch vehicle (EELV) programme, consisting of both Atlas V and Delta IV launch vehicles, aimed to provide the United States with assured access to space, and in time, to reduce the overall cost of launches. The programme has cost around USD 30 billion so far, but while the EELV programme has generally been successful in meeting its assured access to space and cost-saving objectives, according to the US Government Accountability Office (GAO), the programme faces increasing costs, notably because of necessary launcher upgrades to meet government demand and the lack of a strong commercial market (GAO, 2004).

While these factors may explain why the cost of access to space is high, they do not necessarily explain why costs have not declined over time as the technology has improved. One reason is clearly the formidable technological difficulty of developing a reusable launch vehicle, i.e. costs have not declined by much, if at all, because the necessary technological breakthroughs have not been made. Another contributing factor may be the organisational structure of the industry and the very high barriers to entry that currently exist. Because the industry is highly concentrated and its markets are protected for strategic reasons, incentives to innovate may be weaker than in a more open market environment. In particular, the Schumpeterian process of "creative destruction", which is the main driver of innovation in market economies and plays a key role when a paradigm shift is needed to make a breakthrough, does not really occur in the space sector or is weaker than in sectors where entry is easier. In most cases the main motivation for developing launchers is a strategic one, i.e. to achieve independent access to space. The easiest and safest way to do so is often to duplicate what others have done rather than to innovate.

Despite the strong barriers to entry that prevail in the industry, some daring entrepreneurs are nevertheless attempting to challenge incumbents. The entry of low-cost launchers such as the Space Exploration Technology (Space X) Falcon-1 and Falcon-5, offered at USD 6 million and USD 12 million respectively, may indeed represent a major competitive threat for established launcher manufacturers. For instance, under present pricing models, the Falcon-5 launcher may be offered by Space X for up to 70% less than the cost for Boeing's Delta II and Delta IV mediums. As well, Falcon-1 could be a fierce competitor for the new European Vega launcher, which is expected to cost three times as much (Edwards, 2005).

Whether these efforts succeed remains to be seen. Moreover, even if they do, the cost of access to space will remain high. This has two important economic consequences. First, there is a premium on putting a payload into space. As a result, only (weightless) information space applications have been economically viable up to now, although valuable R&D efforts have used the shuttle, Mir and even some re-entry capsules. Second, because launching costs are high, there is a strong incentive to make satellites last as long as possible. However, this raises the problem of obsolescence when space systems compete with terrestrial systems. One solution, which is far from perfect, has sometimes been to put all the intelligence on the ground (e.g. the "bent pipe" approach in telecommunications) or to use software upgrades.

Reducing the cost of access to space would have clear advantages. Some of the potential benefits are obvious: first, space agencies (and more generally developers of space applications) could do more with existing budgets; second, if cost reductions are large enough, they could lead to the development of new applications.

As noted above, overcoming the technological hurdle to reducing the cost of access to space will require breakthroughs in a number of enabling technologies, notably propulsion. This will call, in turn, for substantial and sustained R&D efforts over long periods of time. However, because of security and strategic considerations, such efforts are more often than not carried out at national level and inevitably involve a fair amount of duplication. One possible path to more effective R&D efforts might be pre-competitive R&D collaboration at international level.[23] In this regard, the role of SEMATECH in the semiconductor industry might offer interesting lessons (Box 3.3).

Questions were raised in the early years of SEMATECH regarding the role of the US government, because the venture involved the co-investment of public and private funds in a privately owned and operated consortium and could have been construed as "targeting" (Teece, 1991). However, as noted in Box 3.3, the issue was resolved after 1996 when government funding was discontinued. SEMATECH also increased its international dimension after 1995 through the active participation of companies from Europe and Asia.

One may question the applicability of the SEMATECH model to the space sector on the grounds that international co-operation on sensitive enabling space transport technologies has often been discouraged for strategic reasons. However, SEMATECH has also dealt with sensitive technologies. Moreover, there are some interesting precedents for international co-operation in the space sector. For instance, Boeing's Rocketdyne Propulsion and Power unit and Mitsubishi Heavy Industries of Japan have worked together since 1999 on the

> Box 3.3. **International co operation in semiconductors through SEMATECH**
>
> SEMATECH traces its history back to 1986, when the idea of launching a bold experiment in industry-government co-operation was conceived to strengthen the US semiconductor industry. The consortium, called SEMATECH (SEmiconductor MAnufacturing TECHnology), was formed in 1987, when 14 semiconductor manufacturers based in the United States and the US government came together to solve common manufacturing problems by leveraging resources and sharing risks. Austin, Texas, was chosen as the site, and SEMATECH officially began operations in 1988, with a focus on improving the industry infrastructure, particularly by working with domestic equipment suppliers to improve their capabilities.
>
> By 1994, it was clear that the US semiconductor industry – both device makers and suppliers – had regained strength and market share; at that time, the SEMATECH Board of Directors voted to seek an end to matching federal funding after 1996, reasoning that the industry had returned to health and should no longer receive government support. SEMATECH continued to serve its membership, and the semiconductor industry at large, through advanced technology development in areas such as lithography, front-end processes and interconnect, and through its interactions on manufacturing challenges with an increasingly global supplier base.
>
> SEMATECH's role is to tackle those challenges and to ensure the timely availability of the materials, tools and technology needed by its member companies to stay on the International Technology Roadmap for Semiconductors (ITRS), a plan to make the industry 1 000 times more productive in 15 years than it is today.
>
> *Source:* SEMATECH (2004).

design and development of the MB-XX engine, a new liquid oxygen/liquid hydrogen upper-stage engine for the next generation of expendable satellite launch vehicles. In 2002, they successfully completed the preliminary MB-XX full-scale combustion chamber/injector assembly test programme, and they plan to test the engine jointly in 2005. This co-operation was made possible by working with "black boxes", with each company keeping some elements from the other to respect, in particular, US technology transfer regulations (Ferster, 2004). This would suggest that ways can be found to foster international co-operation for the development of enabling technologies, even when such technologies are deemed to be sensitive.

On balance, a major collaborative pre-competitive R&D effort at international level might be worth exploring further as a way to address more effectively the problem of the high cost of access to space, one of the most vexing problems faced by the space sector and one that has not been solved at national level.

Other significant supply-side features of space applications

While the high cost of access to space is probably the single most important general factor affecting the economic viability of space applications across the board, other supply-side features of these applications also play an important role.

Long lead times

Although some space markets, in the telecommunications sector in particular, are well established,[24] market risks for new space systems are considerable, since their market potential needs to be assessed long before it can be actually tested, owing to the long lead time involved in their development. When these uncertainties are compounded by unexpected (or underestimated) progress in terrestrial technologies (e.g. mobile phones, fibre optics), spectacular failures may follow.

Iridium, a satellite mobile telephony company, is an interesting case in point. In the 1990s, the then-innovative Iridium faced a long lead time to develop the system and secure funding. This put Iridium in the difficult position of lagging behind highly competitive terrestrial services (i.e. rapidly expanding mobile telephony networks), and it was forced to declare bankruptcy in 2000, only two years after it was launched.[25] Iridium clearly illustrates that space actors face high commercial risks (i.e. the market may no longer exist or may have been taken over by another technology when the system becomes operational) and high financial risks (i.e. the investment needs to be made upfront), and have little or no salvage value at the end of the life of the hardware.

Since the costs and risks involved in developing and launching space assets are so high, commercial space applications tend to be deployed only if, at the planning stage, potential investors perceive them to provide a unique and very valuable service or offer considerable advantages over competing terrestrial technologies. However, such precautions may not be enough. Painful experiences like the Iridium debacle have taught investors to be very careful indeed.

To reduce such risks, some firms have attempted to develop cheaper and less complex systems that can be deployed more quickly. For instance, to meet the needs of the marketplace, companies such as Surrey Satellite Technology Ltd. develop small satellites rapidly at lower cost. An increasing number of start-up firms, often small and medium-sized enterprises (SMEs), have used this approach in the launching and satellite manufacturing business since the late 1990s.

Innovative and streamlined operations and improved management efficiencies, linked with new knowledge management techniques and advances in information technologies, lead to some cost savings. However, although entrepreneurs are trying out new ideas or re-using proven systems with new technologies in subsystems, it often takes experienced teams years to master all of the necessary competencies. In other words, space systems still take time to be developed and operated, but the entrepreneurship approach appears to be starting to impact the space sector by dynamising the development of new systems and their downstream applications.

Giving a greater role to SMEs might spur the emergence of new ideas and the development of new products and systems (Lebeau, 2004). The innovative drive of SMEs, their flexibility and their effectiveness make them valuable partners in space projects. This has induced space agencies to establish special programmes to tap into the potential of SMEs.[26]

Economies of scale

Economies of scale play a key role, in both the upstream and the downstream segment of the industry. Upstream firms face high fixed costs because of the importance of R&D activities in the development of space systems. There is therefore a strong tendency towards concentration in this industry segment. The tendency is exacerbated by strategic considerations that tend to fragment markets. As a result, several families of launchers co-exist, each in a quasi-monopoly position in its protected public market, while all attempt to compete for the limited open commercial market. In these circumstances, the price charged to buyers of launch vehicles does not necessarily reflect costs. For example, the cost of developing launchers may not be taken into account, if the launcher was developed with public funds or if the manufacturer received directly or indirectly substantial public subsidies.[27]

Currently, seven families of launchers compete for a rather small market (i.e. Atlas V, Proton, Sea Launch, Delta IV, Ariane 5, H-2A, Long March), which is unlikely to expand much in the decade to come.[28] "Mutual back-up" co-operative mechanisms between launch providers, such as the Launch Services Alliance (LSA), which includes Arianespace, Boeing Launch Systems and Mitsubishi Heavy Industries, give customers greater assurance that their payload will get into orbit, even if their original launch provider defaults. This type of mutual back-up agreement could soon also be applied to government payloads. However, although overall service might be improved for the final user, this type of agreement will not help increase competition and lower costs.

Large economies of scale also prevail downstream because, in most cases, the operation of space systems involves high front-end costs and low marginal costs. For instance, in the satellite communications market the cost of serving an extra user in a given geographical area is practically nil. Economies of scale also prevail in Earth observation, since the cost of producing an extra image, once the system is operational, is very small. This is the case as well for space-based navigation systems. Indeed, a global navigation system could in theory serve any number of clients at no extra cost.

It follows that the larger the market, the more likely it is that space applications may be economically viable. Therefore, space applications will typically be vulnerable to regulations that tend to fragment markets and can greatly benefit from liberalisation efforts (*e.g.* satellite communications have been able to take advantage of a recent World Trade Organisation agreement, which has in effect allowed satellite operators to expand their market across international borders). Another consequence of large economies of scale is that, in any given market, there is a strong tendency towards increased concentration. While such concentration may raise competition policy issues in some markets, it may also be in the public interest if the operators face stiff competition from terrestrial technology.

Dual use

Space technology is by nature "dual use", *i.e.* it can be used both for civil and military purposes. This has both advantages and drawbacks for the development of civil public and commercial applications: on the one hand, the strategic interest of governments in space motivates the development of new systems, which may have useful civil and commercial spin-offs; on the other hand, the same strategic considerations may induce governments to interfere with the activities of private actors, restricting their ability to export or to seek partners at the international level.

Historically, the development of space assets for strategic reasons has been an important source of technological spin-offs for the development of civil applications, both public and private (*e.g.* commercial high-resolution imagery). This is particularly true in the United States, which has a relatively large budget devoted to military space. This trend is likely to continue. For instance, as occurred in the 1980s with the Strategic Defense Initiative, the development of anti-ballistic missiles systems in the 2010s could have tremendous implications for the development of space activities in general. Other governments in Asia and Russia are also developing their military space systems, with potentially significant civil and commercial spin-offs.

The situation is somewhat different in Europe and the rest of the world, where military space plays a much more limited role, although space-related security concerns are on the rise.[29] Because European military space budgets are relatively small, dual use is seen as a cost-effective way for security applications to "free ride" on civil ones (*e.g.* current discussions concerning the foreseen dual use of the Global Monitoring for Environment and Security programme and the use of Galileo by European military forces).[30]

The desire of space-faring nations to have an independent capability to deploy and operate space assets deemed strategically important means that they will support the activities of national firms providing such assets and services, whether these activities are economically viable or not. At the same time, however, international movements of capital, which may lead to changes in the control of these firms, tend to be restricted, while the export of sensitive technologies, products and services are subjected to stringent export controls (Box 3.4).[31] This stifles competition, fragments markets and prevents the efficient allocation of resources.

Following the end of the cold war, the number of actors involved in the development of commercially available, dual-use space products and services has risen dramatically (*e.g.* Russia, Japan, India, China). Although this situation has spurred competition in terms of prices for some specific systems (*e.g.* space launchers), some earlier actors have deplored the practices of some of their competitors, especially in non-market economies (which have clear competitive advantages in terms of human labour costs).

When looking towards the future, an important policy challenge from an overall economic perspective will be to make certain that the dual-use characteristics of space technology are used so as to ensure that security requirements are met in a cost-effective manner, that potential civil and commercial spin-offs are fully exploited, and that the operation of markets is not unduly distorted by strategic considerations.

Application-specific economic strengths and weaknesses

In addition to the general economic strengths and weaknesses described above, each area of application faces particular economic challenges and offers specific opportunities on the demand side.

Space communications

While the commercial future of satellite-based services (including HDTV and iTV) seems assured, the ability of space communications to overcome their weakness in two-way communication, notably satellite broadband services, will very much depend on the policies adopted by governments and the evolution of space and terrestrial technologies. If broadband is perceived exclusively as a commercial activity, satellite broadband will likely play only a relatively minor role.

> **Box 3.4. The Missile Technology Control Regime (MTCR)**
>
> Originally established in 1987, with the aim of controlling exports of missiles capable of delivering weapons of mass destruction, as well as related equipment and technology, the Missile Technology Control Regime (MTCR) is a voluntary arrangement among 34 countries* to restrict the proliferation of ballistic missiles and related technology.
>
> - The MTCR rests on adherence to common export policy guidelines (the MTCR Guidelines) applied to an integral common list of controlled items (the MTCR Equipment, Software and Technology Annex). Many dual use space components (i.e. civil and military applications), including commercially produced ones, are included.
>
> - MTCR partners apply voluntarily the guidelines in their national export controls procedures (*e.g.* International Traffic in Arms Regulations [ITAR] regime in the United States). Generally, partners notify each other of their denials of export licences for specific items to third countries and take into consideration denials by regime partners when reviewing licence applications.
>
> - As the MTCR regime has been increasing its membership (from seven members in 1987 to 34 in 2004), several non-member states have announced they would abide by MTCR Guidelines, as part of their national export control regimes (*e.g.* China entered a second round of negotiations with MTCR in summer 2004).
>
> - While the MTCR has achieved some success in stemming the spread of missile technology, with its principles of enforced technology denial, it has also had the effect of inhibiting the development of some civil and commercial space activities (i.e. restrictive technology transfers).
>
> * Argentina (1993), Australia (1990), Austria (1991), Belgium (1990), Bulgaria (2004), Brazil (1995), Canada (1987), Czech Republic (1998), Denmark (1990), Finland (1991), France (1987), Germany (1987), Greece (1992), Hungary (1993), Iceland (1993), Ireland (1992), Italy (1987), Japan (1987), Luxembourg (1990), Netherlands (1990), New Zealand (1991), Norway (1990), Poland (1998), Portugal (1992), Korea (2001), Russian Federation (1995), South Africa (1995), Spain (1990), Sweden (1991), Switzerland (1992), Turkey (1997), Ukraine (1998), United Kingdom (1987), United States (1987).
> *Source:* Based on MTCR (2004).

There are some major uncertainties regarding the future of satellite broadband. Some important projects originally scheduled for 2004 have been delayed or abandoned. At the beginning of 2005, the iPSTAR satellite, an ambitious Asian initiative, is still on the ground. In the United States – an important test market for the rest of the world – WildBlue will not start commercial service until the second quarter of 2005. Another rather ominous sign is the decision of DirecTV, the largest provider of DTH services in the United States, to use its Spaceway 1 and 2 satellites (now under development) for HDTV instead of the business broadband service they were originally designed to provide. In doing so, the company is writing down the value of the Spaceway assets by as much as USD 1.5 billion. This strongly suggests that DirecTV does not perceive satellite broadband as an attractive business proposition (de Selding, 2005b).

According to Northern Sky Research, a maximum of 3 million rural households and SOHO (small offices and home offices) in North America and 2 5 million in Europe might eventually purchase satellite broadband services if they are made available to them. However, satellite broadband operators could face fierce competition from emerging terrestrial technologies, such as WiMax (see Box 3.5). Tapping the much larger market of satellite TV subscribers would be much more challenging (Northern Sky Research, 2004).

The terms of the equation might change if governments were to assign high value to broadband as a way to extend e-government to all and to contribute to territorial development. In this more favourable public policy context, a case might be made for supporting space-based solutions, if no other technology can provide the same service as effectively, thereby enhancing the potential development of satellite broadband.[32]

From a regulatory perspective, satellite broadband could also be useful as a check on the monopoly position of incumbent terrestrial telecommunications operators in low-density markets where new terrestrial entrants are not present (new terrestrial entrants typically tend to avoid low-density areas and to restrict their offerings to more profitable densely populated areas). However, the danger of such a policy is that the gain achieved thanks to public support may be only temporary as terrestrial technologies gradually catch up.

Earth observation

In the field of Earth observation, the economic problem is somewhat different. The potential value of EO products and services has long been well recognised. However, despite rapid progress over the years, many actual and potential users find that these products and services fall short of expectations or present major stumbling blocks to their effective use. Hence, the market for such products and services has remained small when compared to the cost of developing space assets. Moreover, demand is largely public and competition from terrestrial technology is strong in some segments of the market.

Problems for developing EO products and services have been discussed in the context of the formulation of the European Global Monitoring for Environment and Security (GMES) initiative. In this regard, Brachet (2004)

Box 3.5. **WiMax: a new disruptive technology?**

WiMax is a new wireless technology that could have a major impact not only on providers of satellite broadband services, but more generally on the wireless market at large, including operators of mobile phone services. WiMax, which is backed by a number of large companies, such as Intel, Nokia and AT&T, offers the promise of blanket wireless Internet coverage up to 30 miles from the base station. Initially, WiMax is expected to be a fixed technology that offers compatibility between different vendors' fixed wireless broadband equipment. This should help expand the market for fixed wireless Internet access and enable people in rural areas to access the Internet simply by fixing a WiMax receiver to the outside of their home and plugging it into a WiFi station or directly into a PC. WiMax could also be used as a wireless "last mile" in the developing world since it can carry voice calls using voice-over Internet protocol (VoIP). Instead of laying copper cables, network operators would set up far less expensive WiMax towers, and then install WiMax telephones in subscribers' homes. Internet access could also be provided.

If WiMax can be scaled down to fit inside mobile devices, it can be installed in laptop computers starting in 2006-07. By then, a mobile version of the WiMax standard is expected to have been approved. This could make economically feasible fee-based WiFi-like coverage over wide areas for mobile users. It may also allow operators of WiMax networks to become mobile phone operators by using mobile WiMax in mobile phones.

The technology is still experimental; the first WiMax devices are not expected on the market before the end of 2005. It will be uneconomical compared to cable and DSL in urban areas. It will also be too expensive for use in the developing world, at least for the time being, since early WiMax access devices (which must be fixed to the outside of a building) will cost around USD 500; other forms of wireless links, such as mobile-phone networks, will remain a cheaper way to connect up remote villages. However, it might be a serious competitor for satellite broadband in rural areas. Moreover, the technology could take off with the development of mobile WiMax. Intel plans to start marketing WiMax chips for laptops in late 2006, and the chips are expected to consume only 10% more power than today's WiFi chips. If the technology develops as planned, cheap, mass-produced WiMax chips could allow Intel to dominate the mobile device market in the future, just as it currently dominates the PC market.

Source: The Economist (2004, 2005).

notes that developing an information flow that satisfies the requirements of a European policy on environment and security faces at least three problems:
- Lack of co-ordination among organisations involved in data collection and information production.
- Inadequate intercalibration of the data being collected, both over time and from one source to another.
- Insufficient dialogue between information users and providers.

To remedy this situation, a European shared information service based on a partnership between the main European actors and on a permanent dialogue between stakeholders has been proposed. The European Union is providing financing for 2004-08. It was also noted that to implement an effective GMES strategy by creating an observation system and information production network that fully meets the needs of European policies, it is vital not to limit GMES to its space-based component. An end-to-end approach integrating *in situ* and remotely collected data, data assimilation and modelling techniques, research and long-term monitoring should allow Europe to participate fully in a future comprehensive co-ordinated EO strategy as proposed by the Earth Observation Summit held in Washington, DC, on 31 July 2003.

Another economic challenge facing the development of Earth observation is the fact that, as was observed during the third phase of the project, different economic models are used for data generation and distribution, raising level playing field issues at the international level. This issue is addressed in Chapter 4.

Navigation

The development of space-based navigation applications is likely to expand considerably in the coming years, as the quality of the available signal improves in terms of accuracy and reliability, and as the cost of terminal equipment declines further (EC, 2004). As a consequence, society will become much more dependent on these applications and may require more diversified solutions, such as different levels of precision and reliability for the different application possibilities (*e.g.* highly reliable safety-of-life signal for "life-and-death situations", "commercial" or "guaranteed" signal for services where an interruption would be very damaging or costly to business users).

Civil applications on the public side (*e.g.* air traffic control, road management) appear promising, and some public road transport applications are already successful (*e.g.* monitoring of trucks in transit through Switzerland) (EC, 2003a). On the commercial side, both "telematics" and location-based services are likely to generate increasing sales, but business models are not yet fully established. The profits are currently in the inexpensive consumer equipment. Moreover, it appears that most of the profits are in sales of software, not in sales of hardware. However, possible competition from terrestrial alternatives needs to be taken into account.

The opportunities for a commercial GNSS seem to exist and deserve to be explored further (GMO, 2003), even if is still not clear whether signal operators attempting to provide signal on a commercial basis will be able to capture a sufficiently large segment of the revenue stream to justify their investment, given the "free-rider" problem created by the existence of a free signal. The market entry of Galileo, followed by the future GPS III could pave the way, through their interoperability, to the creation of an open international environment for the development of navigation systems in which competition should drive down costs, to the benefit of end users.[33]

Applications derived from space transport

As noted above, most applications derived from space transport are unlikely to become economically viable in the foreseeable future because: i) the cost of access to space is high and is not expected to decline much; and ii) declines in cost would have to be very substantial (i.e. at least one order of magnitude) to justify private investment in space transport applications.

This may not apply to all space transport applications. As was suggested in the studies carried out in the third phase of the project, suborbital tourism/adventure may attract the attention of space entrepreneurs. The successful flight of SpaceShipOne, winner of the USD 10 million Ansari X-Prize has proven that spaceflight is no longer the exclusive domain of large government programmes and may open the door in the coming years to the creation of space tourism firms. Already, British entrepreneur Richard Branson has started a company called Virgin Galactic to develop the world's first privately funded spaceships to carry commercial passengers on suborbital space flights. Construction is to start in 2005 on the spaceship VSS Enterprise (based on a larger version of SpaceShipOne), with the goal of carrying paying passengers aloft starting in 2007 (Edwards, 2005).

Critics point out that suborbital flight is hardly leading-edge; it has been done for over 40 years. Many of the designs adopted by space entrepreneurs are modelled after earlier experimental "X" craft planes. It is also noted that private suborbital vehicles are cheaper only in that they build on years of R&D paid for by the government. Moreover, there is a big difference between suborbital and orbital flight since the energy needed for the latter is 80 to 100 times that needed to complete a suborbital mission.

However, space tourism advocates argue that the business approach adopted by space entrepreneurs brings a welcome focus on cost control and value for money which may have been lacking in public programmes. They also point out that taking advantage of public R&D is not new: most, if not all, commercial applications in the past have relied heavily on R&D financed by government. It is indeed the only way commercial applications have been able to emerge. Moreover, they note that it is reasonable to expect that the experience acquired with suborbital flights could be of use in the development of a true RLV, even though the main effort in this regard will probably be carried out in other quarters (*e.g.* military space). As was pointed out in the second phase of the project, such an RLV might possibly emerge by 2025 and be competitive with ELVs by 2030.

The role of framework conditions

Whether the technological and economic hurdles reviewed above can be overcome will very much depend on the ability of space actors to focus their efforts effectively, to use resources efficiently and minimise wasteful duplication, to co-operate when it is worthwhile and to devote sufficient attention to the development of innovative solutions. This, in turn will be largely determined by framework conditions, *i.e.* on who the different actors are, how responsibilities are allocated among them and whether the rules of the game that govern their activities provide the right type of incentives.

The economic significance of framework conditions

Several decades of work at the OECD have been devoted to gaining a better understanding of the significance of framework conditions both for specific sectors of the economy and for the economy as a whole. From the outset, the analysis has pointed to the damage that barriers to trade and investment inflict on the economy and to the huge economic benefits generated by an open international system of trade and investment and a level playing field.

In the 1980s, considerable emphasis was placed on the importance of policies that support a stable yet flexible economic environment in which innovation and entrepreneurship can flourish. Subsequently, the Organisation's attention increasingly turned to structural issues, the role of markets and the importance of regulation. It has conducted work on a number of sectors that are important for overall economic performance, beginning with networked industries such as telecommunications and continuing through utilities (*i.e.* electricity generation, gas, water, transport) and on to retail distribution, banking, health and other services. This body of work had made clear the importance of allowing for more private-sector provision of goods and services when it can do a better job than the state; of putting in place a competitive environment that not only helps bring down costs but also stimulates creativity and innovation; of reducing regulatory barriers that are clearly unfriendly to business; and of making those economic activities that remain in the state's domain more efficient and more cost-effective through appropriate changes in laws, regulations and organisational structures.

The OECD's analyses also point to the substantial benefits that can be reaped through greater convergence of standards, nationally and internationally, in sectors as diverse as information and communication technology (ICT), environment and agriculture. Moreover, where the time is not yet ripe or the activity too complex to allow for standards, much can be gained from less binding agreements in the form of guidelines, as in the case of OECD guidelines on such matters as data security, hazardous goods or private pensions. By the same token, the Organisation's work has highlighted the benefits of other forms of institutional co-operation, both among national entities and across borders. This is well illustrated in the field of pre-competitive R&D and technological and scientific innovation.

The significance of framework conditions for space: evidence from case studies conducted in the third phase of the project

The importance of framework conditions for the successful development of space applications – whether public or private – was confirmed in the case studies conducted in the third phase of the project (see Annex A). An interesting finding is the presence of significant commonalities across applications with respect to critical success factors and issues. First, the significance of maintaining a stable and predictable environment was evident in all case studies. Another strong message is the need to deal effectively with uncertainties that relate to liability, notably for emerging applications, as well as the importance of creating and preserving a balanced competitive environment when the services provided by the applications under consideration have to compete with the services offered by other space and non-space actors.

Equitable access to services was another major theme. It extends beyond the digital divide between rural and urban dwellers to encompass questions of equal treatment of individual and national entities regarding access to information and knowledge derived from space activities in general. Moreover, in most case studies, issues related to the generation, distribution and use of information also played a prominent role, notably those relating to intellectual property, the pricing of data and the problem of data confidentiality and privacy.

The case studies also demonstrated that greater compatibility of technological systems, standards, licensing practices and so on are keys to the future development of space applications. Another recurring issue is the central role of infrastructure and the extent to which public authorities should be involved in their provision and operation. Finally, in a number of instances there was a clear-cut case to be made for encouraging government support of R&D.

Many of these critical issues fall squarely under the responsibility of governments. Moreover, they extend far beyond the traditional field of space policy and need to be considered in a much broader policy context (*e.g.* economic, social and environmental policies).

Framework conditions clearly matter for the space sector. They will be covered in Chapters 4 and 5 and appear as two distinct but linked categories: institutional conditions and legal and regulatory conditions.

Notes

1. Fixed satellite services (FSS) include all communications services that occur when a signal is sent to a given position (*i.e.* a "fixed" station). FSS represent the bulk of satellite communications, including very different services such as basic telephone communications or television broadcasting (see direct broadcast satellite services below). Mobile satellite services (MSS) are provided by networks of communications satellites intended for use with mobile and portable devices on land, air or sea (*e.g.* satellite mobile telephones).

2. See Chapter 2 for more details on direct broadcast satellite (DBS) developments.

3. In North America alone, sales of GPS equipment in 2003 are estimated at between USD 3.4 billion (Frost and Sullivan, 2003) and USD 4.7 billion, with asset tracking and fleet management accounting for USD 670 million of that total (Bates, 2003).

4. The entry into service of Galileo may in fact be delayed. While European Commission documents continue to refer to Galileo's in-service date as 2008, industry and government officials agree that the constellation is unlikely to be ready for commercial use before 2011 (de Selding, 2005a, p. 4).

5. These GMO estimates are based on the assumption that compatible Galileo-GPS chipsets will be available. It should be noted that numbers provided by governmental agencies and consulting firms tend to offer an idea of markets trends, rather than absolute market values.

6. The United States has a long list of failed programmes such as the X-30, X-33, X-34, the second generation RLV and the Orbital Space Plane (OSP). In Europe, limited efforts have been made at national level. One promising example is the Phoenix RLV prototype built by EADS Space Transport in Germany which could be selected as one of Europe's future means of space access under ESA's Future Launcher Preparatory Programme (Edwards, 2005).

7. RLV concepts vary widely (*i.e.* fully reusable vs. partly reusable, vertical take-off vs. plane-like takeoff). Many largely conceptual studies have been conducted by agencies and industry for the past 20 years (*e.g.* NASA, 1994).

8. Propulsion is a crucial enabler of future capabilities of launchers and spacecraft (*e.g.* Earth observation satellites). In launch vehicles, propulsion accounts for 70-90% of vehicle weight and 40-60% of system costs (Kelly, 2004). A satellite's lifespan is limited to the lesser of either power or propulsion life, whence the importance of developing smaller, lighter, more powerful and more affordable propulsion and power systems.

9. In late 2004, the US Defense Advanced Research Projects Agency (DARPA) awarded contracts, ranging in value from USD 8 million to USD 11.7 million, to four companies seeking to meet the government's demand for a Force Application and Launch from

Continental US (FALCON) small launch vehicle. The companies – AirLaunch LLC, Lockheed Martin, Microcosm, and SpaceX – are in various stages of progress on the project. SpaceX is perceived by some analysts to be furthest along and is set to perform an "early, responsive launch demonstration" in 2005 (FAA, 2004).

10. Early Bird represented a major step in intercontinental communications. When launched in 1965, it provided almost ten times the capacity of submarine telephone cables for almost one-tenth the price. Satellite remained competitive with cable until the advent of optical fibres in the late 1980s. From 1965 to the late 1980s, communications cost to consumers went from over USD 10 per minute to less than USD 1 per minute in current dollars. The decrease is even more impressive when the effect of inflation is included.

11. However, in the 2004 hurricanes in Florida, satellite communications proved less vulnerable than cable to severe weather conditions, inducing cable subscribers to switch to satellite. For instance, Mediacom, a major US cable operator, reported a loss of basic cable subscribers during the third quarter of 2004, which it blamed on the effects of the severe hurricane season and more pressure from satellite TV competition. Mediacom said it lost an estimated 8 000 subscribers to Hurricane Ivan, which affected company operations in Alabama, Florida and Mississippi. According to Mediacom, the hurricane initially disrupted cable service to more than 100 000 basic subscribers in the three states (SkyReport, 2004a).

12. Forecasts for the United States suggest that more than 14 million HDTV receivers may have been sold by the end of 2004, while in Japan some 10 million are expected to be on the market by the end of 2006 (SES Astra, 2004).

13. For instance, in 2002, SES Astra brought to market the Broad-Band Interactive System (BBI), at that time the first commercial use of the Ku-band spectrum, to forward multimedia information via geostationary satellite to small dishes, and of the Ka-band spectrum to receive in return similar information or to transfer files using the same satellite and dishes (SSPI, 2004). Important research is also being carried out by manufacturers of space systems. For instance, Alenia Spazio's research on satellite broadband solutions includes the development of new digital platforms, based on its SkyPlex turbo transponder, which allows signals sent from different ground stations to be regenerated and aggregated directly on board the satellite, while the use of the Ka band allows a reduction in the size of the terminal antenna.

14. In the United States, a number of phone companies intend to roll out fibre-based services to consumers so as to provide them with "triple play", i.e. digital video, broadband and phone services. The companies involved so far include SBC Communications, Verizon and Bell South. For instance, in November 2004, SBC outlined plans to overbuild 17 million homes with fibre-to-the-node technology and deployment of fibre-to-the-premises (FTTP) services to another 1 million by 2007. However, financial analysts tend to be sceptical about the company's ability to take a substantial share of the video market – a market that is already saturated and highly competitive – without deep discounts that reduce the profitability of the investment (SkyReport, 2004b).

15. Examples of X-band EO systems include the COSMO-SkyMed satellites, with the first to be launched in 2005 (Alenia Spazio, 2004).

16. The ISS is the largest platform in space, with modules that are built in different countries and connected for the first time in space. In fact, space systems are more and more using plug-and-play subsystems that may be interchangeable (e.g. satellite buses adaptable to different launchers). This has been affecting all space applications and will increasingly do so if international standards and interfaces are developed in parallel.

3. SUPPLY CONDITIONS: STRENGTHS AND WEAKNESSES OF SPACE

17. In the aftermath of the successful completion of the X-Prize competition, Bigelow Aerospace is now launching a new prize, the "America's Space Prize". The winners of this USD 50 million prize will have to build a spacecraft capable of taking a crew of no less than five people to an altitude of 400 km and complete two orbits of the Earth at that altitude. They will have to repeat the same accomplishment within 60 days, with real passengers. Moreover, no more than 20% of the spacecraft's hardware can be expandable. The spacecraft must also demonstrate the ability to dock with Bigelow Aerospace's inflatable space habitat and be able to stay docked in orbit for up to six months (David, 2004).

18. Although the primary goal of a possible robotic mission could have been to install a deorbit module on the Hubble Space Telescope, NASA was also studying the feasibility of performing other tasks, such as installing new batteries, gyros and possibly scientific instruments that would have enhanced the observatory's ability to peer even more deeply into the universe (NASA, Johnson Space Center, 2004).

19. Another more practical reason, according to NASA officials, is that a human mission aboard the shuttle is likely to come too late for Hubble. Even if the first shuttle flight occurs in spring 2005, as tentatively planned, it would be unrealistic to expect that after a few test flights – in which problems might be discovered and cause further delays – that Hubble could be serviced before its batteries or gyros fail. A human journey to Hubble would be at least fifth on the return-to-flight priority list, behind shakeout flights and at least one trip to the ISS (Britt, 2004).

20. The primary objective of ESA's Aurora programme is to create, and then implement, a European long-term plan for the robotic and human exploration of the solar system, with Mars, the Moon and the asteroids as the most likely targets. The current preparatory phase of the Aurora Exploration Programme is to culminate in a full programme proposal, which will be submitted to the next ESA Council meeting at ministerial level, currently scheduled for the end of 2005 (ESA, 2004b).

21. Moreover, because production runs are short, the production process tends to be very labour-intensive. The launcher manufacturing industry is in fact a "prototype" industry where productions runs are rarely large enough to justify substantial investment in process automation.

22. Changes in R&D priorities also increase the overall costs of developing launchers, with no significant benefits. NASA successively gave up, for instance, its research on a single stage to orbit (SSTO) vehicle, as well as on a two stages to orbit (TSTO) system it was working on until late 2002. The ensuing Orbital Space Plane (OSP) project, labelled the "American Soyuz" at an estimated cost of USD 13 billion to USD 18 billion, was also abandoned.

23. "Pre-competitive research" here refers to R&D that is distant from the market and focused on generic or enabling technologies rather than technologies targeted at particular markets. Typically, the research effort is not expected to produce commercially usable technologies or products, but rather to reach the stage of demonstrating feasibility or providing research prototypes.

24. According to the *Financial Times*, satellite operators are considered as "boring" by financial analysts because they generate predictable cash flows over long periods of time (*Financial Times*, 17 August 2004).

25. Originally, by providing communications to anyone anywhere, Iridium expected 5 million subscribers and revenues of USD 600 million a year. The original investors lost USD 5 billion in the venture. By December 2000, a group of new investors acquired the original assets (including 66 satellites) for USD 25 million, a very small fraction of the cost of the system, and launched a new company, Iridium Satellite LLC. The company received a USD 72 million contract from the Pentagon for secure voice

communications, a contract that could be worth USD 252 million through 2007 if all options are exercised. Other clients include the UK Defence Ministry, Colombia's national police and the government of Alberta. Ironically, things are looking up for the company now, although on a much smaller scale than once envisaged: revenues were up 44% in 2003 over 2002, and the satellite fleet may remain operational until 2014, eight years longer than originally expected.

26. For instance, it is estimated that in Europe some 200 SMEs (with workforces of 100 or less) fill in design and supply niches that complement the activities of the bigger players. These SMEs tend to cluster around three areas: small systems integration work, production of space equipment, and software, engineering and research. Recognising the potential of SMEs, ESA, which devotes 90% of its budget to contracts with European industry, established in the late 1990s the "SME Initiative" to enable ESA and the European space industry to tap into the potential of leading SMEs and to open opportunities for them to work more intensively with ESA and space contractors (ESA, 2004d).

27. If the government pays for the R&D (at billions of USD a year), it is charged as a sunk cost which does not become part of the accounting system for future government use of a vehicle, which is pretty much the marginal cost of a current operating budget. On the other hand, private companies must recover their investments so that the price charged by a private firm will be much higher.

28. The short-term demand for the geostationary orbit (36 000 km), the primary destination of commercial satellites, is generally expected to be only around 15-20 satellites a year, with an increasing number of satellites needed by the end of the decade for replacement and new services.

29. In this regard, Logsdon (2002) notes that despite current efforts, the development of a European security capacity faces two important obstacles: the tension between national and European levels from a sovereignty perspective and divergence regarding co-operation with the United States.

30. Overall, military space budgets in Europe amount to only about 5% of US expenditures in this area. European firms have long complained that this puts them at a competitive disadvantage vis-à-vis their American counterparts. The European Commission's White Paper on Space (EC, 2003b) recommended that member states develop common dual-use space systems and set up a panel of experts in the field of space and security (SPASEC) to examine how to move forward. The Panel began work in 2004 and is currently considering management options for identifying, maintaining and updating operational requirements for pan-European space and security capabilities.

31. For instance, in most space-faring countries governments can control the activities of commercial EO firms during times of national security emergencies and for other strategic purposes.

32. For instance, Cohendet et al. (2005) note that the deployment of satellite broadband in low density areas could allow rapid implementation of broadband services which are essential for the development of activities offering strong externalities, such as distance education, telehealth and e-government.

33. The Russian GLONASS navigation system could also become an important player in the future. Three new spacecraft were launched in December 2004, bringing the number of GLONASS spacecraft in service to 11. The Russians plan to increase the constellation to at least 18 by 2007. It should have 24 satellites in its final configuration. A preliminary agreement was also signed in December 2004 with the United States to promote GLONASS compatibility and interoperability with GPS for worldwide civil use (*Aviation Week and Space Technology*, 10 January 2005).

Bibliography

Alenia Spazio (2004), "Alenia Spazio Signs COSMO-Skymed Contract", *Press release*, 22 December.

Ball Aerospace (2004), "Orbital Express Advanced Technology Demonstration Program", *www.ball.com/aerospace/oexpress.html*, accessed 20 September 2004.

Bates, J. (2003), "Security Concerns Boosting Sales of GPS Devices", *Space News*, 30 June.

Berger, B. (2003), "NASA Proposes $300 Million Tug To Deorbit Hubble", *Space News*, 24 November.

Brachet, G. (2004), "From Initial Ideas to a European Plan: GMES as an Exemplar of European Space Strategy", *Space Policy*, Vol. 20, pp. 7-15.

Britt, Robert Roy (2004), "Details Emerge in Robotic Plan to Service Hubble", *Space.com*, 4 May.

Cohendet, P. and L. Stojak (2005), "La fracture numérique en Europe : Les enjeux économiques et sociaux au regard d'une 'Europe de la connaissance'", *Futuribles*, No. 305, February.

David, L. (2004), "Rules set for $50 Million Space Prize", *Space News*, 8 November.

EC – European Commission (2003a), "Business in Satellite Navigation: An Overview of Market Developments and Emerging Applications", March, Brussels.

EC (2003b), *White Paper – Space: A New European Frontier for an Expanding Union: An Action Plan for Implementing the European Space Policy*, European Commission, Brussels.

EC (2004), "Inception Study to Support the Development of a Business Plan for the GALILEO Programme: Executive Summary Phase II", Brussels.

Edwards, J. (2005), "New Dawn for RLVs", *Aviation Week and Space Technology*, 15 January.

ESA – European Space Agency (2004a), "Robots – Our Helpers in Space", *Press Release*, 29 November.

ESA (2004b), "ESA's Exploration Programme 'Aurora' gets Further Boost", *Press Release*, 20 December.

ESA (2004c), Automated Transfer Vehicle (ATV), *www.esa.int/export/SPECIALS/ATV/*, accessed 13 August 2004.

ESA (2004d), *The ESA SME Initiative*, ESA, Paris.

Euroconsult (2004), *World Prospects for Government Space Markets: 2004 Edition*, Paris.

FAA – Federal Aviation Administration (2004b), *Commercial Space Transportation: Quarterly Launch Report, Fourth Quarter 2004*, Washington, DC.

Ferster, W. (2004), "Visualizing a Future in Propulsion", *Space News*, 24 August.

GAO – Government Accountability Office (2004), "Defense Space Activities: Continuation of Evolved Expendable Launch Vehicle Program's Progress to Date Subject to Some Uncertainty", GAO-04-778R, 24 June.

GMO – Galileo Market Observatory (2003), "Satellite Navigation Market Intelligence Briefing, European Commission GALILEI Project", GALI-ESYS-DD112 v4.1.1, May.

ICSU – International Council for Science (2004), "Priority Area Assessment on Scientific Data and Information", Draft Report, August.

JAXA – Japan Aerospace Exploration Agency (2004), Wideband Internetworking Engineering Test and Demonstration Satellite (WINDS), *www.jaxa.jp/missions/ projects/sat/tsushin/winds/index_e.html*, accessed 7 September 2004.

Kelly, Michael F. (2004), "Powering the Future", *Air and Space Power Journal*, Spring, *www.airpower.maxwell.af.mil/airchronicles/apj/apj04/spr04/kelly.html*, accessed 5 August.

Lawler, A. (2004), "Stormy Forecast for Climate Science", *Science*, Vol. 305, 20 August.

Lebeau, A. (2004), "Promouvoir la petite enterprise en Europe: Les enseignements du Small Business Act américain", *Futuribles*, 303, December.

Logsdon, J. (2002), "A Security Space Capability for Europe? Implications for US Policy", *Space Policy*, Vol. 18, pp. 271 – 280.

MTCR – Missile Technology Control Regime (2004), General Presentation, *www.mtcr.info/*, accessed 10 August 2004.

NASA – National Aeronautics and Space Administration (1994), "Commercial Space Transportation Study (CSTS)", conducted by Boeing, May, *www.hq.nasa.gov/ webaccess/CommSpaceTrans/*, accessed 3 October 2004.

NASA Johnson Space Center (2004), "Human-like NASA Space Robot Goes Mobile with Leg, Wheels", *Press Release*, 6 August.

NASA – National Aeronautics and Space Administration Marshall Space Flight Centre (2004), "NASA Factsheet: DART Demonstrator to Test Future Autonomous Rendezvous Technologies in Orbit", FS-2004-08-113-MSFC, September.

Northern Sky Research (2004), "WildBlue Moves Closer to Commercial Launch, An Industry Status Briefing from Northern Sky Research", Report, January.

de Selding, P. (2005a), "Final Bids Are In To Run Galileo's Navigation System", *Space News*, 31 January.

de Selding, P. (2005b), "DirecTV Revenues Soar in 2004 but Operating Costs High", *Space News*, 31 January.

SEMATECH (2004), *www.sematech.org/corporate/index.htm*, accessed 20 September 2004.

SES Astra (2004), "Astra Launches Its HDTV Demo Channel", Press release, 1 September.

SkyReport (2004a), "Cable Weathers 3Q Storm", SkyReport online, 12 November.

SkyReport (2004b), "Wall Street Takes Notice of Telco Fiber Plans", SkyReport online, 16 November.

SSPI – Society of Satellite Professionals International (2004), Industry Innovator Awards Program, *www.sspi.org/*, accessed September 2004.

Teece, David J. (1991), "Support Policies for Strategic Industries: Impact on Home Economies", *Strategic Industries in a Global Economy: Policy Issues for the 1990s*, OECD International Futures Programme, OECD, Paris.

The Economist (2004), "Wi-Fi's Big Brother", 11 March.

The Economist (2005), "World Domination Postponed", 27 January.

US DOT – United States Department of Transportation (2001), "Vulnerability Assessment of the Transportation Infrastructure Relying on the Global Positioning System", Report prepared for the DOT by the Volpe National Transportation Systems Center, 29 August.

ISBN 92-64-00832-2
Space 2030
Tackling Society's Challenges
© OECD 2005

Chapter 4

Framework Conditions: Institutional Aspects

> *Framework conditions (i.e. the existing institutional, legal and regulatory regime) largely determine how society is organised for meeting future challenges. It is therefore important to assess whether such conditions will encourage the development of the space systems that can be expected to help solve enduring socio-economic challenges. This chapter focuses on institutional aspects (i.e. who does what, and does the current situation needs to be changed?), while legal and regulatory aspects are addressed in Chapter 5. The institutional issues addressed here first include the role of space agencies, their position in the overall machinery of government, their links (if any) with the military and questions raised by the setting up of major international projects. The balance of the chapter is devoted to institutional issues raised by the operation of space applications. Each main area of application (telecommunications, Earth observation, navigation) is considered in turn as pertinent institutional solutions differ significantly across applications, notably regarding the role to be played by public and private actors.*

4. FRAMEWORK CONDITIONS: INSTITUTIONAL ASPECTS

Introduction

The institutional framework for space activities varies from country to country. It is largely shaped by the general policy objectives of decision makers, by the role they seek to play on the world scene, by the importance they attach to space for fulfilling their objectives, by the relative value they assign to different space activities (military, civil, commercial) and by their views regarding the role of public and private actors.

Historically, strategic objectives have largely driven space developments, although countries have differed in terms of their objectives and thus in terms of their activities. There are, as a result, significant differences among them, and these are reflected in large differences in the size of their space budgets and in the way these budgets are allocated among space activities. For instance, US leaders spend more on space and apparently attach more importance to space for achieving their objectives than their European counterparts. Moreover, their strategic use of space has a much greater military dimension, while civil space activities are primarily viewed as tools for developing technological skills and achieving leadership, although economic aspects are also well recognised. By contrast, Europeans give greater relative importance to civil space, with industrial development as a major objective (through exploration and application-oriented programmes).

Despite understandable differences, major space-faring countries have generally adopted a similar broad institutional model for conducting their space-related activities. This "generic" model involves three general sets of actors: i) public agencies that focus on space research and development (R&D), typically space agencies; ii) public and/or private agencies responsible for the operation of space systems and the development of downstream applications; and iii) public and/or private organisations responsible for the upstream segment of the space industry (i.e. spacecraft and launcher manufacturers, providers of launching services).

The main features of the model are as follows:

- Public bodies focusing largely on upstream space activities (i.e. "space agencies" or "space administrations" in major space-faring countries) are established to carry out basic research and develop space applications. (*Rationale*: To achieve results, a critical mass of expertise must be developed over long periods of time. Moreover this R&D work has a high investment threshold which only the public sector can afford and justify on the basis of public good criteria, i.e. because it fulfils a demand for a public good.)

- Once an application has been developed to the demonstration stage, operational agencies are typically set up to run the application. (*Rationale*: Space agencies need to focus on their R&D functions and are not equipped to provide services on an ongoing basis to a broad range of different customers, although they can continue to provide technical support to operational agencies.)

- Operational agencies may be run as purely public bodies, financed by the state (*e.g.* the European Organisation for the Exploitation of Meteorological Satellites [EUMETSAT] in Europe or the National Oceanic and Atmospheric Administration [NOAA] in the United States for meteorological satellites). They may also be run on a "commercial" basis, i.e. generate a substantial share, if not all, of their revenue from the sale of services. (*Rationale*: Operational agencies should logically remain public when the service they provide has a strong public good dimension; they should be operated on a commercial basis if the service they provide is a private good.)

- Among commercial operators, some may still receive some level of state support or be partially owned by the state, depending on market conditions (*e.g.* the French supplier of satellite data Spot Image); others may start as public actors and eventually be privatised (*e.g.* the telecommunications firms Eutelsat, Intelsat); still others may operate on a purely private basis from the start (*e.g.* SES Global as a private start-up in the 1980s). (*Rationale*: Private actors are generally best placed to run applications that can be operated on a commercial basis and that generate enough revenue to be economically viable.)

- Private actors in the upstream segment of the space industry provide input to space agencies and co-operate with them to develop new systems and construct basic components. (*Rationale*: Private actors have complementary expertise that space agencies lack and can take over production activities, which are beyond the R&D mandate of space agencies, once the new system is developed.)

- Private actors also play a key role in identifying and seizing new business opportunities that take advantage of the technologies developed in co-operation with or by space agencies, especially in the downstream segment, where innovative applications may only involve a small – albeit essential – space segment. (*Rationale*: This contributes to the full exploitation of the research efforts; private actors are best placed to do this effectively.)

The general model outlined above offers only a very rough guide to institutional arrangements for space activities, and one that is not universally applied. However, it provides a useful starting point for considering the institutional issues facing major space actors. The following discussion first considers issues related to the role of space agencies (broadly defined) and then turns to issues related to the operation of space applications in the three major fields of application of space technology: telecommunications, Earth observation and navigation. An applications-oriented approach is necessary to fully take into account the significant differences among these applications, on both the supply and the demand sides.

Issues related to the role of space agencies

Strictly speaking, not all space-faring countries have a dedicated space agency; however, they all have, at some level, an authority dealing with space R&D activities. In what follows, the term "space agencies" is used for convenience to include all public bodies in charge of conducting basic R&D activities and developing space applications.

The activities of space agencies raise three sets of issues from an institutional point of view. The first relates to the main focus of space programmes: What should be their objectives and what priorities should be established among them? The second pertains to the organisational structure: Given the objectives to be pursued, where should the agency fit in the overall machinery of government? A related question, taking into account the dual use nature of space technology, is to determine the relationship (if any) between the space and military agencies. Finally, a third set of issues relates to international co-operation. Since countries' space agencies pursue similar objectives, what is the scope for pooling resources and expertise to take full account of synergies without compromising security and sovereignty objectives?

The focus of space agencies

Most space-faring countries engage in internal discussions regarding what the roles of space agencies and other space-related organisations should be and the type of impact their actions have on the overall space sector. Historically, space agencies have played a key role in supporting and conducting R&D and scientific programmes, being at first directly involved in the development and running of the programmes, and then increasingly supporting the development of an indigenous space industry by contracting out projects. Space agencies basically have three main missions:

- The development of space technology (*e.g.* basic research in propulsion, development of new launchers, satellites, Earth stations).
- The use of space for scientific missions (*e.g.* space exploration, environmental research).
- The development of space applications (including some that may become commercial).

A central question involves the resources to be allocated to each of these areas of activity. On the one hand, there are strong pressures to ensure that space programmes generate significant, tangible and highly visible payoffs. On the other, there is a danger that too much emphasis on short-term returns may undermine the necessary longer-term efforts that only public agencies are in a position to carry out (*e.g.* basic research needed to reduce the cost of access to space) and discourage private entrepreneurs from investing in space ventures.

The problem arises notably when the activity is already commercially mature, as in the case of satellite communications. In this regard, one of the most difficult issues is the appropriate role of national governments in developing new satellite technology and systems. There are several schools of thought on this subject.

- One school holds that satellite communications have become commercially viable and that industry should now be expected to finance the development of the technology needed to succeed in the 21st century.
- A second school holds that space communications is the only truly successful space enterprise and that public investments should go to areas promising the greatest payoff and thus help spur the next big breakthroughs in satellite technology and systems.
- Finally, a third school argues that private funding can develop the commercial technology, but for emergency and public services such as health, education, etc., special systems technology to fill specific niches may make sense.

In telecommunications, Korea and Japan are investing in the most rapidly growing markets and generally reflect the second school of thought. India, China, Canada and Brazil are shaping space technology to meet public social needs, as in the third school of thought, but are hoping for future commercial pay-offs as well.

Regarding the development of satellite broadband, the third school of thought seems to be the most widespread, as in Canada (*e.g.* development of the Anik F2 satellite) and in France (*e.g.* Agora project). The United States, which seems largely to follow the first school of thought, may be a major exception. However, the National Aeronautics and Space Administration (NASA) has been very active in the development of satellite broadband technology in the past, and the US Department of Defense (DoD) continues to carry out major efforts on satellite communications, and these may have civil and commercial applications.

For other space applications, the approaches adopted by governments vary considerably. For instance, the second school of thought tends to dominate in Europe regarding the development of Galileo, as well as strategic/sovereignty considerations, while the Global Navigation Satellite System (GNSS) remains

4. FRAMEWORK CONDITIONS: INSTITUTIONAL ASPECTS

essentially military in the United States. In the field of Earth observation, the third school of thought prevails, although attempts are made to put at least some segments on a commercial footing.

A similar issue arises for space tourism. Some argue strongly that it should be left entirely to the private sector. Others point out, quite convincingly, that a modest level of public support might facilitate the emergence of a suborbital space tourism industry in the coming years, generating jobs and value added, to the extent that demand exists for its services. Moreover, the industry could contribute over time to the development of cheaper and more reliable space vehicles. Annex A provides more details on the formulation of possible business models for space tourism.

The organisational structure

The organisational structure adopted for a space agency is largely determined by the importance attached by the government to space-related activities: the greater the importance attached to space, the more likely it is that the space agency will report to a central agency of government or be a fully independent agency reporting directly to the prime minister or the president and the larger the budget allocated to space is likely to be. However, this is not always the case.[1]

The organisational structure will also depend on the main mandates given to the space agency. If the main focus of activities is on science, the agency should logically report to the ministry of science and technology. If on the other hand, the main emphasis is on commercial applications, it is more likely to report to economic ministries. Because they typically receive a multiple mandate, space agencies often tend to report to several ministries (including the ministry of defence if they engage in military space activities). This may be a source of difficulties for formulating a space agency's priorities.

It is interesting to see how these general principles apply in practice in major space-faring nations. The first approach has the agency or administration report directly to the executive branch of government (*e.g.* president or prime minister):

- In the United States, NASA is an independent administration within the US government, although it has to go through the Office of Management and Budget (OMB) and Congress for funding.[2] The US Department of Defense, as an independent Cabinet agency, has a less visible but very substantial space programme.[3]
- In India, the Department of Space is an independent body at the same level as other ministries, directly under the prime minister's office. It is co-ordinated with the inter-ministerial Indian Space Commission and supervises the activities of the Indian Space Research Organisation (ISRO), established in 1969. The model has the advantage of providing, at least in

theory, a direct, formal and visible link between developers of space systems on the one hand and governmental user departments and policy makers on the other.

The second approach, the most widespread in space-faring countries, creates a dedicated agency supervised by one or more ministries, traditionally a ministry of research or economy.

- France's space agency (Centre National d'Études Spatiales [CNES]) has been supervised jointly by the ministries of research and of defence since 1997.
- In Korea, the Korea Aerospace Research Institute (KARI), established in 1989, is supervised by the Ministry of Science and Technology.
- The Japan Aerospace Exploration Agency (JAXA), established in October 2003, has merged the activities of three former agencies (the Institute of Space and Aeronautical Science [ISAS], the National Aerospace Laboratory [NAL] and the National Space Development Agency [NASDA]). It functions under the authority of different ministries, in particular the Ministry of Education, Culture, Sports, Science and Technology (MEXT) and the Ministry of Public Management, Home Affairs, Posts and Telecommunications.
- In the Netherlands, the agency for aerospace programmes (NIVR) reports to the Ministry of Economic Affairs, but various other ministries are involved in the selection of the agency's management team.
- In Sweden, the Swedish National Space Board (*Rymdstyrelsen*), is a central governmental agency under the Ministry of Industry, Employment and Communication.
- In Italy, the Italian Space Agency (*Agenzia Spaziale Italiana* [ASI]), established in 1988, reports to the Ministry of the Universities and of Scientific and Technological Research.

A third approach, typically adopted when space budgets are modest, establishes a dedicated space organisation with several proactive governmental stakeholders.

- In the United Kingdom, the British National Space Centre (BNSC), located in the Department of Trade and Industry, is a voluntary partnership, formed among ten government departments and research councils, to co-ordinate British civil space activities.

A fourth approach merges space activities with other activities to develop an integrated national R&D structure.

- In Belgium, the Federal Science Policy Office, under the Ministry for Economy, Energy, Foreign Trade and Science Policy, is responsible for the management of Belgian public participation in space-related R&D activities, primarily within the framework of international organisations to which

Belgium belongs (*e.g.* ESA, EUMETSAT) or within the framework of bilateral agreements (*e.g.* with France on the SPOT satellites).[4]

- In Austria, recent organisational changes take a similar approach. The Austrian Research Promotion Agency (FFG – Österreichische Forschungsförderungsgesellschaft) was established in September 2004 as the main player in Austrian public funding for R&D (Unkart, 2004). This new agency has merged four different organisations, including the Austrian Space Agency which was created in 1972, and is to be supervised by the Federal Ministry of Transport, Innovation and Technology and the Federal Ministry of Economy and Labour.[5]

Finally, creating an intergovernmental organisation (IGO) can be extremely useful to promote international co-operation and avoid duplication of effort. In this regard, the European Space Agency (ESA) is an interesting case in point. It was established in 1975 as an IGO with a mission to provide and promote the exploitation of space science, research, technology and applications for the benefit of its member states.[6]

In recognition of the importance of space for Europe as a whole, ESA and the European Union have established a Framework Agreement for the formulation of an overall European Space Policy, as well as for establishing a common basis and appropriate practical arrangements for co-operation between the two institutions. This agreement was adopted in 2003 and entered into force in May 2004. To implement it, a Space Council has been established at ministerial level to co-ordinate and facilitate co-operative activities between the two bodies. The Space Council met for the first time in November 2004. It offered ministers representing EU and ESA member states a first opportunity to discuss jointly the development of a coherent overall European space programme. This programme – which is expected to be endorsed by the Space Council at the end of 2005 – will constitute a common platform for achieving the objectives set by the European Space Policy. Another important recent development is the decision taken by the European Commission in late 2004 to shift responsibility for its space activities from the Research Commissioner to a new Directorate on Enterprise and Industry.

Table 4.1 provides details on current institutional arrangements a number of space-faring countries.

As the stakeholders in space programmes now include communities beyond those of scientists and R&D actors (*i.e.* industrial sectors, defence), the positioning of space agencies and related operational agencies in the complex machinery of governments is increasingly called into question in many countries. For instance, the position of France's CNES is regularly discussed, since it has both public and private characteristics as an industrial and commercial public body (*établissement public industriel et commercial* – EPIC). The

CNES is a key stakeholder in a number of French and European commercial firms involved in space transport, Earth observation and satellite-based location and data collection.[7] Some analysts recommend that it should be shifted to the Ministry of Trade and Industry, rather than stay under the supervision of the Ministry of Research (Conseil Économique et Social, 2003). A similar discussion is also taking place in Italy. The Italian Parliament is currently discussing the possible shift of ASI from the research ministry to either an independent ministry for space and innovation, or directly under the prime minister's office, with the creation of an inter-ministerial space committee (Taverna and Nativa, 2004). In Norway, the Norwegian Space Centre, established as a foundation in 1987, became in 2004 a governmental agency under the Ministry of Trade and Industry (NRS, 2004).[8] Finally, the transfer of responsibility for space activities at the European Commission from the Research Commissioner to a new Directorate on Enterprise and Industry reflects the greater importance given to the pursuit of industrial policy objectives and to the development of space-based services and infrastructure at European level.

Although the situation varies from country to country, the general trend seems to be towards giving greater importance to space agencies' links to other governmental agencies, as the range of space applications expands, and to give greater importance to highly visible activities. However, higher visibility is not always reflected in higher budgets. Moreover, these moves are seen by some observers as running the risk of losing crucial support from scientists by too great a focus on short-term objectives, thereby overlooking important governmental obligations with regard to research in basic science and technology.

What links with the military?

This publication deals mainly with civil aspects of space activities, but space technologies are by nature dual, and military developments have often paved the way for the development of civil and commercial applications. Moreover, there is a gradual blurring between civil and military security, as they often use similar tools (for instance an Earth observation satellite can be used to support civil disaster management efforts or military operations).

In some countries, the budgets for defence R&D tend to outweigh spending on civil space, and policy makers increasingly depend on space assets for all types of defence and security missions. The growing role of space in military affairs was recognised by the Rumsfeld Commission (January 2001), which noted the present extent of US dependence on space, the rapid pace at which this dependence is increasing, and the vulnerabilities it creates. The Commission further noted that meeting the national security needs of the 21st century would require far-reaching reorganisation.[9]

4. FRAMEWORK CONDITIONS: INSTITUTIONAL ASPECTS

Table 4.1. **Space institutions and lines of reporting**

	Cabinet-level body or Executive		Economy, Industry Ministry
	Research, Science Ministry		Member states (international organisation)

	Space institution	Reports to	Comments
Asia	China National Space Administration (CNSA), an administration in the Commission of Science, Technology and Industry for National Defence (COSTIND), a ministry-level agency	State Council	Established in 1993, at the same time as the China Aerospace Corporation (CASC), which deals with commercial activities. China has several other organisations involved in space activities: the China Satellite Launch and Tracking General (CLTC), the State Science and Technology Commission (SSTC), the China Academy of Science, the China Academy of Engineering, and the Ministry of Information Industry (MII)
	Indian Space Research Organisation (ISRO)	Department of Space, directly under the prime minister's office	Established in 1969. Space policy is co-ordinated with the inter-ministerial Indian Space Commission
	Japan Aerospace Exploration Agency (JAXA)	Ministry of Education, Culture, Sports, Science and Technology (MEXT), and Ministry of Public Management, Home Affairs, Posts and Telecommunications	Established in October 2003, merging the activities of three former agencies: the Institute of Space and Aeronautical Science (ISAS), the National Aerospace Laboratory (NAL), and the National Space Development Agency (NASDA)
	Korea Aerospace Research Institute (KARI)	Ministry of Science and Technology	Established in 1989
North America	Canadian Space Agency (CSA)	Ministry of Industry	Established in 1989
	National Aeronautics and Space Administration (NASA)	President	Established in 1958. Since 1996, the National Science and Technology Council (NSTC) is the main forum for executive R&D policy making, though the National Security Council (NSC) co-chairs some of the policy processes. The US Congress allocates the budget annually, and the OMB supervises the budget
	National Oceanic and Atmospheric Administration (NOAA)	Secretary of Commerce (a member of the US President's Cabinet)	Established in 1970. In 1982, creation of NOAA's Satellites and Information Service, also known as the National Environmental Satellite, Data and Information Service (NESDIS)
South America	Argentina's *Comisión Nacional de Actividades Espaciales* (CONAE)	President	Established in 1991
	Agencia Espacial Brasileira (AEB) or Brazilian Space Agency	Executive Office of the President	Established in 1994
Europe	Austrian Research Promotion Agency (FFG – *Österreichische Forschungsförderungsgesellschaft*)	Federal Ministry of Transport, Innovation and Technology, and Federal Ministry of Economy and Labour	Established in 2004, key player in Austrian public funding for R&D. The agency merged four different organisations, including the Austrian Space Agency created in 1972
	Belgium Federal Science Policy Office	Ministry for Economy, Energy, Foreign Trade and Science Policy	Since 1975, successive science policy offices involved in space activities

4. FRAMEWORK CONDITIONS: INSTITUTIONAL ASPECTS

Table 4.1. **Space institutions and lines of reporting** (cont.)

	Cabinet-level body or Executive		Economy, Industry Ministry
	Research, Science Ministry		Member states (international organisation)

	Space institution	Reports to	Comments
Europe (cont.)	British National Space Centre (BNSC)	The BNSC is owned and managed by the ten BNSC partners, and headquartered in the Department of Trade and Industry (DTI)	Established in 1985 as an *ad hoc* interdepartmental working arrangement – partnership of ten government departments and research councils (*e.g.* Office of Science and Technology, Department of Trade and Industry, Ministry of Defence)
	Centre National d'Études Spatiales (CNES)	Ministry of Research and Defence Ministry	Established in 1961, joint ministerial supervision with Defence since 1997
	Deutsches Zentrum für Luft- und Raumfahrt (DLR) Germany's Aerospace Centre	German Ministry for Education and Research	First established in 1969; aside from the Ministry for Education and Research, DLR activities are funded by different sources, depending on the project: transport ministry, defence ministry, and by contributions from the *Länder* (federal states)
	Italy's *Agenzia Spaziale Italiana*	Ministry of the Universities and of Scientific and Technological Research	Established in 1988, the agency co-operates with numerous entities active in space technology and with the president of the Council of Ministers
	Netherlands Agency for Aerospace Programmes (NIVR)	Ministry of Economic Affairs	In 1971, the Netherlands Agency for Aircraft Development (NIV), created in 1947, added space to its activities and became the Netherlands Agency for Aerospace Programmes: NIVR; many ministries are involved in the agency's management: Economic Affairs, Transport, Public Works and Water Management, Education, Culture and Science, Defence and Finance
	Norwegian Space Centre	Ministry of Trade and Industry	The Norwegian Space Centre, established in 1987 as a foundation, became in 2004 a governmental agency under the Ministry of Trade and Industry
	Swiss Space Office (SSO)	State Secretary for Science and Research	Works in co-operation with the advisory Federal Space Affairs Commission (CFAS) composed of representatives of scientific, industrial and user circles and the inter-ministerial Committee for Space Affairs (IKAR)
	Russian Aviation and Space Agency (RAKA)	President	Established in 1992, acts in co-ordination with the Ministry of Defence for some activities
	National Space Agency of Ukraine	President	Established as an independent administration in 1992.
	European Space Agency	15 member states (intergovernmental organisation)	Established in 1975, by the merging of the European Space Research Organisation (ESRO) and the European Launcher Development Organisation (ELDO), both established in 1962
	European Organization for the Exploitation of Meteorological Satellites (EUMETSAT)	18 member states and 9 co-operating states (intergovernmental organisation)	Established in 1986, it is responsible for the launch and operations of European meteorological satellites and for delivering data to end users, as well as contributing to the operational monitoring of climate and the detection of global climate changes

Source: Agency Web sites and Verger (2002).

For some experts, the links between civil space-related organisations and their military counterparts is becoming a key issue. Some recommend enhanced synergies between the civil and military agencies, in order to rationalise R&D budgets and foster innovation. The institutional framework for such co-operation already exists in some countries. In France for example, CNES missions include leading and implementing the country's military space programme, in partnership with the military and the French defence procurement agency, the *Délégation Générale pour l'Armement* (DGA). The co-operation is steered by the space co-ordination group (*Groupe de Coordination Espace*), chaired by the head of the military joint-chiefs of staff.[10]

In the United States, the DoD and NASA have co-operated closely ever since the creation of NASA, although the agency was specifically and purposely formed as an open, civil administration with a mandate to disseminate all scientific information to the public (its research results are mostly available and not classified).

When NASA was formed, many DoD centres of excellence were transferred to it to serve as its core. The Mercury and Gemini missions, for instance, all flew on DoD launch vehicles and the two agencies have continued to collaborate across the full spectrum of space activities, including launch, communications, sensors, materials and life sciences. Both depend on rockets manufactured by private-sector contractors to launch the payloads – such as spy satellites, weather satellites or scientific instruments – that are required for national security or for carrying out research in space. Hence, both agencies could co-ordinate their development and purchasing of rockets with a view to increasing the reliability and lower the cost of launch vehicles. Their combined efforts might also encourage, among other things, the emergence of new, entrepreneurial companies able to launch payloads into space.[11] Co-operation is also active in other areas, notably the development of weather satellites.[12]

In other countries, mixing civil and military efforts is not always self-evident. As an intergovernmental organisation, ESA has since its inception developed technologies that are inherently dual-use, but according to its Convention it can only promote co-operation for "exclusively peaceful purposes". Recent years have seen a consensus among its member states that "peaceful purposes" do not exclude co-operation on security and defence programmes of a non-aggressive type (ESA, 2003b). ESA is thus developing links to other European security-related organisations with an interest in space, such as the new European Defence Agency (EDA).

Space agencies and international co-operation

To the extent that space administrations are involved in the production of public goods or that the goals they pursue are shared by other agencies, there is a strong incentive for such agencies to co-operate. Indeed, co-operation offers a number of advantages:

- By pooling resources the agencies can undertake projects that would be beyond their means to carry out on their own.
- By combining their knowledge and skills, they can take advantage of a pool of expertise larger than each is able to tap in house.
- By working with others engaged in related activities, each agency may be able to influence the decisions other agencies are taking.
- By promoting greater international co-operation, they may also be able to foster international good will in general and strengthen their "soft power" at international level.

On balance, international co-operation in the space sector has often been quite successful in the past, especially for scientific missions (*e.g.* an agency's scientific instruments carried on board a foreign satellite to multiply measurement possibilities), but it has not always been without problems. Space has above all a strategic and geopolitical dimension, often with military implications. Decisions relating to international space programmes are often highly political, owing to the difficulties inherent in accommodating conflicting national interests.

The European approach

The creation of ESA established an original and effective co-operative system for orchestrating space co-operation at European level. Through ESA's programmes, western European countries have developed communications, weather, scientific, Earth observation and technology demonstration satellites and deep space probes. ESA's successful creation of the Ariane series of launch vehicles is one of the key achievements of this international collaboration.[13] Further efforts are under way to rationalise and harmonise even more technology developments by actors throughout Europe through technology road mapping (see Box 4.1).

In order to involve its member states, ESA has applied since its creation a principle of geographical or "fair return" (also called *juste retour*) which has constantly evolved. This principle guarantees member states an industrial return in the form of contracts awarded to their industries, equivalent to their contributions to the Agency's programmes. This is a rather clear incentive for members to commit themselves to regular contributions to different programmes. This fair return policy also encourages smaller partner countries to participate more actively in the programmes, in that it gives them some assurance that their industry will derive some direct or indirect benefit from the space activity they help to finance.[14]

4. FRAMEWORK CONDITIONS: INSTITUTIONAL ASPECTS

> **Box 4.1. The European technology road map harmonisation process**
>
> **The concept of technology road mapping**
>
> A technology road map is a planning process that identifies, evaluates and selects technology alternatives to satisfy a set of specific needs in terms of products or services (i.e. what enabling technologies need to be developed to reach Earth's orbit as cheaply as possible). It is often cost-efficient to harmonise different long-term R&D efforts (as in the European technology harmonisation process).
>
> The main benefit of technology road mapping is that it provides information to help make better decisions for technology investment. It is a difficult exercise, sometimes quite subjective, as diverse opinions on the validity of certain technologies need to be confronted, and budgets have to be spread among disciplines and technologies. When political or corporate priorities change, technology road maps are usually among the first programmatic items to be overhauled.
>
> **European co-ordination efforts**
>
> European efforts to harmonise space technology developments across the continent are relatively recent. The European Space Technology Requirements Document (Dossier 0), first issued in 1999 and updated in 2002, sets technology needs and priorities for space-related R&D for future and potential space activities of European interest within a ten-year horizon. It is aligned with every major application (Earth observation, telecommunications, navigation, science and exploration, human spaceflight and microgravity, spacecraft bus, ground segment, payload data exploitation and engineering tools, space transport, innovative and prospective technologies). Based on these priorities, the European Space Technology Master Plan (ESTMP) compiles different roadmaps for overall technology planning (involving ESA, space agencies, etc.) The ESTMP also includes a Europe-wide directory of space activities, providing space stakeholders across Europe with an overview of European institutions' ongoing and planned technology activities.
>
> *Source:* Based on ESA (2004).

One drawback of the approach is that it may distort competition, fragment production and result in some duplication of effort. The next ESA/EU Space Council, to be held in spring 2005, will look into these questions and investigate what future industrial policy principles should be adopted, as it is to draft an overall European Space Programme. In this regard, it was already pointed out in the *White Paper on Space* (EC, 2003) that European industry would undoubtedly benefit from more flexibility in Europe's approach to space procurement. The White Paper argued that when the objective is to optimise Europe's collective interests, a broader definition of geographical return could be used more creatively so as not to discourage companies from making more cross-border investment, particularly in the new member states, and in order to avoid unwarranted duplication of effort.

More generally, the development of an appropriate institutional framework at European level is particularly challenging. It involves first of all establishing the relationship between ESA and national space agencies so as to avoid gaps and redundancies in their overall activities.[15] Moreover, this involves efforts to combine ESA's "technology push" approach with the European Union's emphasis on "demand pull" for the development of Europe-wide space infrastructure. While the combination is viewed as potentially fruitful, it raises particular challenges in light of the differences between ESA and the EU in terms of membership and working principles. In this regard the *White Paper* envisages a two-phase approach: phase 1 (2004-07): the topics covered by the framework agreement between the EU and ESA are implemented; phase 2 (from 2007), if space is officially recognised as an area of shared competence between the EU and its members as proposed in the European Constitution Treaty, "ESA should then be positioned within the EU framework and its Convention modified accordingly" (EC, 2003).

As clearly demonstrated by European experience in the context of ESA's activities, the setting up of a structured co-operative programme requires extensive and time-consuming negotiations among the major contributors before the programme is effectively launched. This is also true of most major co-operative space efforts, although clear scientific missions have in general been largely easier to set up than large applications-oriented projects (*e.g.* the Galileo negotiations took years longer than at first envisaged).

Once agreed on and launched, a co-operative effort remains very vulnerable, particularly if it is a highly visible, large project. First, because it is in the public eye, it is subjected to intense scrutiny, and problems that arise become the topic of heated debate among "instant experts". Second, if the different participants are not of generally equivalent weight, synergies and emulation among the partners may be stifled. Finally, problems experienced by one partner inevitably affect the others.

The International Space Station and the Joint Striker fighter experience

The International Space Station (ISS) is an interesting case in point. Some authors see it as a great success. For instance, according to former NASA administrator Daniel Goldin (2004), the ISS is not only a technical success but also a shining example of international co-operation (*i.e.* a platform for peace).

It also paves the way to further exploration by increasing our knowledge and experience of working in space. Moreover, it is a laboratory for advancing science and transferring reliable technologies. It is for now the only permanent foothold in space that provides the environment necessary to develop the tools and technologies that will take humans farther into space. Other analysts are more critical. They point out that debates regarding the usefulness and cost of this huge orbital platform have raged in all 16 partner countries, and that the grounding of the space shuttle has cost the space station partners hundreds of millions of dollars. From an institutional perspective, the terms of the intergovernmental agreement (IGA) that governs co-operation among the various parties and sets out their obligations have also been criticised. Several ISS partners, notably in Europe, feel that, from a legal perspective, the terms of the IGA do not put them on an equal footing with their US partner.[16] An important lesson is that capability dependence and an unequal legal basis do not facilitate cost-efficient co-operation (especially for the dependent parties).

A major difficulty for international co-operation is the tension that exists between the collective interest of all parties to use resources as efficiently as possible, on the one hand, and, on the other, the desire of each party to derive maximum benefits for its national actors from the co-operative effort. As noted, ESA tackled the problem by implementing a geographical return provision that ensures that all partners receive industrial benefits that reflect their level of contribution to the joint project. While this approach may satisfy each partner individually and provides an incentive for co-operation, it may not be an efficient way to allocate resources overall if the return provision is applied too rigidly.

A radically different approach was taken for the development of the US Joint Strike fighter (JSF), a non-space approach that may be used as a reference for selected large space projects involving American agencies, if the Aldridge Commission recommendations are followed (Aldrige, 2004).[17] The model, which would probably have to be adapted to the specificities of space sector co-operation (e.g. barter agreements vs. purchases of equipment), takes an integrated approach, in which project participants provide components to a single integrator that manages the project's cost, schedule, performance and risk. In the case of the JSF, foreign and domestic suppliers compete for industrial work in the project under a "best value" approach implemented through American prime contractors. Hence, although the partners' domestic companies may receive multi-billion dollar contracts, there is also a risk that their industry will not get any contracts, as there is no government-guaranteed work-share.[18]

Lessons learned

While the experience acquired in other sectors may provide interesting models for international co-operation on space, lessons drawn from past or ongoing collaborative space efforts are also useful, whether the co-operation was set up to build a large infrastructure or to co-ordinate existing systems.

For instance, the ISS experience has shown that long-term formal space partnerships must be structured so that, if necessary, they can evolve over time (Pryke *et al.*, 2002). This is needed not only because new partners may join (*e.g.* the US invitation to Russia to participate in the ISS in 1993 required extensive renegotiation of the original 1988 agreements and led to the current agreements, signed in 1998), but also because national policies may change dramatically and endanger the entire partnership.

In constructing any co-operative programme, it is also necessary to recognise that each partner has national priorities that must be accommodated. An international partnership brings together willing partners seeking political and economic leverage on their investment. In this regard, the international partnership should be perceived as a "win-win" proposition, *i.e.* as providing tangible benefits to all partners. The partners should also have a similar broad general vision of what is to be achieved, although they may have different interests, expectations and reasons for participating. High-level political leadership may also be necessary to garner international support.

In the case of Earth observation, interesting examples of international co-operation are provided by the Co-ordination Group for Meteorological Satellites (CGMS)[19] or the International Charter for Space and Major Disasters.[20] Both work well, mainly because partners have relatively comparable capabilities and share a common vision (*i.e.* in the case of the CGMS, to gather and diffuse important weather information worldwide; in the case of the Charter, to provide space-acquired data and associated information and services to civil protection agencies worldwide in the event of major disasters) (Brachet, 2004). An important aspect is that in both of these co-ordinating mechanisms, clear links are forged with actual and potential users of space data, and this helps to foster broad public support for the initiatives.

It is clear that institutional solutions need to be tailored to each co-operative effort. The need to find adequate models may become increasingly important in future, as the sheer dimension of the enduring socio-economic challenges that decision makers will face will increasingly require the use of sustainable international space systems.

Issues related to the operation of space applications

Once an application has been developed by a space agency, the question arises of whether it should be run by the space agency or whether a separate entity should be created for this purpose.

The solution usually adopted is to create a separate entity, since an accumulation of operational activities would, over time, overburden the space agency and distract it from its main research mission. The approach taken to creating this entity – as well as its relationship with the space agency – varies from country to country, depending on overall institutional and legal arrangements as well as on the nature of the applications.

In some countries (*e.g.* France, India, Sweden) it is possible to create commercial subsidiaries of the space agency for this purpose, in others (*e.g.* the United States) the entity must be either public or private.[21] In still others, a public-private partnership approach – which may involve public entities other than the space agency – can be adopted.

As the case studies in the third phase of the project demonstrated, one major difficulty at the conceptual level, when considering the status of operating agencies, is that the definitions of some key terms such as "commercial" or "private" are often not clear, and may vary from country to country. For instance, the term "commercial" may have meanings ranging from government enterprise to private-sector activity.

A government enterprise could be said to operate on a commercial basis if it generates most of its revenue from the sale of goods or services to the general public. This applied, for instance, to a company like the automobile manufacturer Renault when it was wholly owned by the French government and competed with private-sector firms. A government enterprise that sells goods and services to final consumers as a public monopoly can also be said to operate on a commercial basis, although the profit motive may be missing. A government enterprise may operate on a partially commercial basis if it derives part of its revenue from the sale of goods to the general public and receives at the same time public funds to complete its budget.[22]

By contrast, privatisation implies the transfer of ownership of assets from the government to a private company, which can address either a business and consumer market or a government market. In the former case, the entrepreneur assumes the business risks of the activity. However, questions arise regarding the status of companies that operate under an anchor tenancy contract or are the main or even the sole provider of critical products to governments. In such cases, the client may end up bearing the risk, even if nominally it is supposed to be borne by the supplier.[23]

The institutional solution adopted for the operating agency varies from application to application, depending on whether the output can be considered a public or a private good. In satellite telecommunications, the output is essentially private, so that a private applications agency is the most appropriate solution. In the case of Earth observation, the output has a strong public good element, so that the applications agency will be essentially public, although a segment of the activity (*e.g.* high resolution imagery) may support private operations. For global navigation systems, for example, the jury is still out regarding the feasibility of private operation.

Strategic considerations also play a central role. First, private operation will be allowed only if it does not interfere with strategic imperatives. Moreover, if private operation of a potentially strategic service is allowed, it will be subject to public scrutiny and may receive public support in one form or another.

Telecommunications

The role of space agencies in telecommunications is currently largely limited to conducting or contributing to R&D activity, while the once publicly owned operating agencies have been privatised over time. Trade liberalisation has contributed significantly to the expansion of the application and its privatisation.

In the 1980s, the telecommunications sector started to attract many new private actors. For instance, in 1985, SES Global, then a small telecommunications start-up, entered the space arena with an initial small investment and a state guarantee. By 2004, the company had become the world's biggest satellite operator, with a fleet of 40 satellites across the globe.

In the 1990s, several intergovernmental satellite operators, with increasingly commercial activities, were privatised successfully (*e.g.* Intelsat, Inmarsat).[24] In the case of Inmarsat, two complementary bodies were formed in April 1999: a limited company, which has since served a broad range of communications markets, and the International Mobile Satellite Organisation, an intergovernmental body established to ensure that Inmarsat continues to meet its public service obligations, in particular those of its Global Maritime Distress and Safety System.

A final constraint on the operation of Inmarsat and Intelsat as private entities was lifted recently by a new US law (S.2896) that basically spares both companies from the need for an initial public offering (IPO) to meet the requirements of the 1999 US ORBIT Act. Such an IPO was considered by some competitors (*e.g.* SES Global) as necessary to ensure transparency for companies that had long been dominant globally, owing to their government-owned status. The case made (convincingly in the eyes of US legislators) by

Inmarsat and Intelsat was that their privatisation and subsequent ownership dilution made them normal, private corporations with little resemblance to the intergovernmental organisations they once were (de Selding, 2004c). A similar process has taken place in Europe with the privatisation of Eutelsat.

The role of public actors in Europe

Despite the privatisation of space communications, space agencies have played in the past and continue to play today a strong supporting role in the development of satellite communications applications, notably for those that are expected to generate substantial societal benefits or have substantial strategic or military uses. One example is the CNES' Agora[25] programme to counter the digital divide. CNES launched the programme in the summer of 2004 with the objective of developing a new satellite broadband telecommunications system, offering three services at competitive costs: high-speed Internet access, interactive television programmes and voice-over-Internet Protocol (VoIP) (CNES, 2004). It proposes to launch a first geostationary satellite in 2007 (to cover France) and a second in 2010 (to cover Europe more generally). While for now CNES plays the central role, Agora is to be privately owned, managed and operated. Negotiations are ongoing with space manufacturers for the provision of a platform and for testing the new dedicated telecommunications instruments to be used. Discussions are also under way with private satellite operators that would be asked to invest in this shared European infrastructure. Regarding the overall value chain, there would be one satellite infrastructure operator (SIO), either an existing operator or a venture of several operators, that would procure and operate the satellites and lease the transponders over a lifetime of 15 years, and one or more multimedia service providers (MSP) per European country, that would procure or lease a gateway set, lease transponders or a share of transponders and develop offers adapted to their customers. Time to market is a central issue, and it is envisaged that a test of service provision could begin in 2005-06 using existing satellites. However, R&D funded by CNES (and ESA) is expected to provide the necessary technologies to produce relatively cheap terminals and gateways from 2006. It is hoped that full-scale development can start in 2005.

The UK Skynet 5 is another interesting example of public involvement in the development of European satellite communications. The choice of a public-private partnership (PPP) model for Skynet 5 is based on two considerations: i) the fact that no commercially available off-the-shelf solution matched the full requirements set out by the UK government; and ii) the realisation that, if a dedicated system were to be developed, part of the capacity created in this way for the non-secure/resilient services, and which was not used by the government, might be subcontracted out to ordinary commercial service providers.

In many respects, the market served by Skynet 5 differs from the commercial broadband market described above, because it consists of a basket of managed services to predefined sites for the UK government and its allies (some highly secure and resilient, and some not), using satellites for service continuity with the military system Skynet 4. However, there is a parallel in that the UK government has contracted for a stated duration and volume of services with particular characteristics, without specifying the technology or commercial model to be used. This allowed Skynet 5 bidders to design a platform supporting UK needs, but allowed them to offer excess capacity to other buyers on commercial terms. The UK government's requirement created the necessary critical mass from the outset for the platform's lifetime.

Therefore, public administrations might use a form of PPP to acquire broadband capacity if they are prepared to aggregate their requirements and allow bidders to use the contract to leverage the wider market. However, many issues may arise, of technology neutrality, competition with existing commercial players, and the legitimacy of demand aggregation (e.g. if an existing small service provider could support local requirements terrestrially more cheaply than a centralised satellite solution), which would have to be addressed.[26]

The role of public actors in North America

In North America, public agencies have played and continue to play a major supporting role for the development of satellite communications. For instance, in the United States, NASA was a main promoter of the Ka-band. It developed notably the Advanced Communications Technology Satellite (ACTS), an experimental satellite that opened Ka-band for space communications and demonstrated the use of narrow spot beam antennas, ultra-wideband transponders (900 MHz), and onboard digital processing and switching. The project supported experiments from industry, academia and the government, and acted as a catalyst for the acceptance and implementation of the technologies used in next-generation satellite systems.

The US DoD has also contributed significantly to the development of satellite communications technologies. A recent example is a new generation of communications satellites to serve tactical users, developed by a team led by Lockheed Martin, which should become operational by 2010. The new Mobile User Objective System (MUOS) is being developed as a replacement to the UHF Follow On constellation. It will provide global satellite communications narrowband connectivity (up to 384 kpbs) for voice, video and data for American and allied military services. MUOS satellites will be fully compatible with the existing UHF Follow On system and associated legacy terminals, while dramatically increasing military communications availability by leveraging third-generation (3G) commercial cellular advances, which are a significant improvement over previous networking technologies. The DoD has also been instrumental in the development of private entities such as Iridium, which serves both military and civil clients.

However, the development of new civil satellite communication applications in the United States has been largely left to the private sector (*e.g.* WildBlue), although some public programmes encourage the deployment of the technology to serve rural and remote areas (*e.g.* FCC support for rural communications). To initiate its broadband service, WildBlue is leasing transponders from Anik F2, a satellite developed for the operator Telesat by the Canadian Space Agency, and which is partially funded by the Canadian government.[27]

Earth observation

As the case studies in the third phase of the project made clear, choosing an appropriate institutional model for developing space applications is more difficult in the case of Earth observation (EO). This is because of the as yet limited private market for EO products and services, and because of the strong public good nature of many such products, notably those with broad area coverage.

The public model

When the Earth observation product or service has strong public good characteristics, strictly public solutions appear to be the most appropriate. For instance, some of the most successful developments concern the operation of meteorological satellites by governmental organisations, such as EUMETSAT, an intergovernmental agency, in Europe and NOAA in the United States, in collaboration with the World Meteorological Organisation (WMO).

EUMETSAT is responsible for the launch and operation of the European meteorology satellites and for delivering data to end users, as well as contributing to the operational monitoring of climate and the detection of global climate changes. EUMETSAT's 18 member states fund the programmes and their respective national meteorological services are the primary users, although a tradition of data exchange among meteorological services allows many non-members access to the meteorological data for the preparation of their forecasts (EUMETSAT, 2003).[28] In the next decade, the Meteosat Second Generation system will become the primary European source of geostationary observations over Europe and Africa. It will be one of the main EUMETSAT contributions to the future Global Observing System of the WMO, which will provide important weather forecasts worldwide (WMO, 2004).

In the United States, NOAA has the dual role of regulating and operating satellites. It issues licences for private EO satellites, carries out R&D on weather satellites and space weather issues, and co-ordinates the meteorological system with the DoD. The public status of weather systems has been confirmed by Congress, which has determined that the system would be operated as a public good, and no attempt at privatising it would be permitted (private companies had submitted proposals).

The failed privatisation of Landsat confirms the pertinence of the public model for broad area coverage and constitutes an instructive saga. As early as 1979, when the Landsat programme was transferred to NOAA, discussions about its commercialisation had begun, at a time when there was much uncertainty about NOAA itself as an organisation. In particular, there were serious doubts about whether NOAA should be located in the Department of Commerce (DoC). As a result, the Landsat programme did not have very high priority at NOAA. Uncertainty and pressures to shift operational costs to the private sector resulted in an acceleration of commercialisation, although studies made at the time (1981-83) suggested that Landsat could not be successfully commercialised without a substantial government subsidy. Indeed, the "privatised" Landsat's attempts to raise prices well above marginal cost to recover start-up costs were unsuccessful. They merely resulted in significant market losses as new entrants (i.e. Spot Image in 1986; the Indian IRS-1 system in 1988) destroyed its monopoly position. As a result, the Landsat programme had to be returned to the federal government by 1992.

The main challenge to be addressed when adopting a public solution is to be able to devise a public model that is sustainable over time and that fully meets the needs of users (ICSU, 2004).

In some cases this is quite easy to do. For instance, for meteorological applications, the case for the creation of a specialised public entity is straightforward: the nature of the service offered is clearly defined and its value is well recognised. Moreover, users can be easily identified, and they co-operate fully with the producer of data in defining successive generations of satellites. Thus, even though the deployment of the second generation of EUMETSAT has just started, plans are already under way to ensure a seamless switchover from the second to the third generation of Meteosat satellites which should take place around 2015. This involves a co-operative effort between EUMETSAT, ESA and EUMETSAT's main customers (i.e. national meteorological services and operational agencies in member states, the European Centre for Medium Range Weather Forecasts and EUMETNET). Consultations, combined with the results of studies undertaken in collaboration with leading experts, formed the basis for defining the Meteosat Third Generation (MTG) programme (EUMETSAT, 2003).

However, the close relationship between data users and data producers that exists in the case of weather applications is the exception rather than the rule. In other areas, such as environmental applications, the customer base is large and diverse, with very different levels of expertise. Another problem is that the source of funding is not secure since space agencies, which largely foot the bill for environmental applications, cannot be expected to fund operating satellites for any length of time beyond the demonstration phase. It follows that a clear and deliberate effort is needed to set up the necessary funding mechanisms. This requires full appreciation of the merit of the activity at the highest level and among the general public. This is the case today for weather satellites, and could also be true for security applications in the coming years. It is less clear for environmental satellites, as the general public has not yet fully apprehended the scope and seriousness of environmental problems and the contribution environmental satellites can make (see Chapter 2 for more details).

The private model

When EO products have private market potential (e.g. high-resolution EO products), a PPP or private solution may be the most appropriate approach. Buying an operational system requires a large investment (on average, from USD 100 million to USD 500 million just for the satellite and its instruments), and several million for annual operation. Governments and commercial operators have been looking at ways to share costs by developing partnership schemes. For example, the SPOT 5 EO satellite cost around EUR 533 million (launch included). To finance the satellite, the firm Spot Image and the French government set up a partnership structure that shares the costs between the operator and the public investor. Canada has tested another type of PPP model for the Radarsat-2 mission, a satellite co-funded by the Canadian Space Agency (CSA) and MacDonald Dettwiler (MDA). MDA will own and operate the Radarsat-2 satellite, which is scheduled for launch in 2005 and will be the next-generation Canadian commercial synthetic aperture radar (SAR) satellite. The CSA's investment should be offset by the imagery provided by MDA at low cost to Canadian government agencies over the satellite's lifetime.

In the United States, a purely private solution was adopted following a review of US commercial Earth observation led by the National Security Council in 2002. It calls for the establishment of a secured government imagery purchase agreement for ensuring control of high-resolution imagery and for providing, at the same time, guaranteed revenue to private EO operators. The programme, called ClearView, was established in January 2003 for this purpose. The programme is run by the US National Geospatial Intelligence Agency (NGA), which has replaced NIMA (the National Imagery and Mapping Agency), and supports the Pentagon and other agencies in procuring commercial satellite imagery. It is a five-year framework contract with a base performance period of three years and two additional one-year options, although a second ClearView programme is in preparation.[29] The programme replaces what NIMA officials considered a cumbersome licensing structure with a single licence that allows timely imagery to be shared with all the agency's potential partners.

While industry would like to see a civil version of the ClearView programme, its implementation by civil agencies is not easy. One problem is that there are no reliable projections of civil spending on satellite imagery, because the relevant agencies do not have specific budgets for this purpose. US civil agencies combined spent less than USD 10 million on commercial satellite imagery in 2003. However, some agencies, notably the US Geological Survey, the Forest Service and NOAA, have been using the existing ClearView contract for imagery purchases, as the ClearView programme contains provisions that allow this. Hence, ClearView could be a viable option, if civil agencies were able to bring their requirements together. However, more widespread civil use of the existing ClearView mechanism might require new licensing agreements that allow for more latitude in sharing of imagery, because civil agencies often work with non-government organisations other than those used by military agencies.

The main problem faced by EO data providers is that the private market has not developed as expected. For instance, nearly 18 years after launching its first satellite with ambitions for stimulating a global private-sector business, Spot Image has concluded that the only EO business worth seeking currently is that of serving government customers; within that market, the military remains by far the most important, according to Spot Image Chairman Jean-Marc Nasr (de Selding, 2003). Indeed, for Nasr, the private commercial market is an illusion. The term "commercial market" really means sales to civil and military customers in governments outside the nation whose taxpayers paid for the observation satellite system. Spot Image expects that commercial private sector demand will remain marginal for the foreseeable future.[30]

In the United States, the main players also continue to depend heavily on public demand. When DigitalGlobe, Orbimage and Space Imaging were planning their first satellites, they expected (like Spot Image) that a large commercial market would provide the bulk of their business. However, this has not happened, in part because the companies took too long to get their product to market (all three suffered launch failures before deploying their first high-resolution imaging satellites). They were also overly optimistic about commercial customers' response to their offerings. Moreover, some largely ignored the commercial market in favour of public users until recently. In effect, government sponsorship actually distracted the companies from pursuing commercial business.

According to Matt O'Connell, the chief executive officer of Orbimage, another obstacle facing Earth observation companies may be cultural, because of the sector's strong technical orientation and its relatively new foray into the commercial arena. In a recent interview, he noted that the main challenge he has had to face since taking over the company in October 2001 is to teach engineers to think like entrepreneurs. Engineers concern themselves with systems that operate perfectly, and they are not always interested in factors such as profitability (Bates, 2004a).

The general view in the industry is that the potential commercial market is large (*e.g.* insurance, urban planning and agriculture), but that it is very segmented and requires EO data providers to serve thousands of customers. Hence, the private market is more competitive and requires business models different from those that apply to public markets. Instead of omnibus multi-year and multimillion dollar contracts under NextView, commercial orders are likely to be far more numerous but scattered and worth only a couple of thousand dollars each.[31] This means that thousands of customers are needed to generate a business worth hundreds of millions of dollars.

It has been estimated that Orbimage and DigitalGlobe have about seven years to build up the commercial market necessary to have a non-US-government funding source to support a third generation of commercial imagery satellites (Bates, 2004b). This is no easy task, although there are encouraging signs: significant gains in commercial business have been achieved recently. For instance, the non-US-government revenues of DigitalGlobe grew by 40% in 2004, and Space Imaging's commercial business is growing by 15-20% a year, although it represents only 10% of total revenue (USD 20 million out of USD 200 million).

In spite of these recent gains, the situation of private EO players remains fragile worldwide. None of the companies providing data on a regular basis has a full in-orbit backup for its satellites. Moreover, failure to be granted a government contract can have a devastating effect on a company.[32]

Navigation

Global navigation satellite systems represent a relatively new and major application of space technology. While GNSS were first developed for military purposes (GPS in the United States, GLONASS in Russia), they have increasingly found civil applications over the years.[33]

In 1984 President Reagan announced that a portion of the capabilities of GPS would be made available to the civil community, and the first GPS civil products appeared in the mid-1980s.[34] Today, GPS has become *de facto* a major domestic and international resource with important civil uses in the United States and throughout the world, although it remains a military system.

The deployment of other GNSS, such as Galileo, creates a new policy environment, given that each of the systems has a global vocation and that the nations with a stake in its deployment and operation may have conflicting policy objectives. If one excludes security concerns, which are beyond the scope of this project, the future development of satellite navigation systems raises two main questions from an institutional perspective:

- How can the existing systems meet the needs of users (and offer interoperability), without involving too much duplication?

- As civil public and private use increases, what role should be given to public and private actors in the development and the operation of GNSS and related systems?

Meeting the needs of users

GNSS have potentially a broad range of civil applications, each with different required navigation performance (RNP) parameters (see Box 4.2). These applications include first, transport applications (aeronautical, maritime, road, rail and pedestrian), each with its own characteristic needs. Second, GNSS have a wide spectrum of engineering uses (*e.g.* monitoring of structures such as bridges or dams) and provide input to geographical information systems (GIS). Third, GNSS provide numerous opportunities for agriculture (*e.g.* precision farming), fisheries (*e.g.* for safety of fishing fleets, but also for supervising the

---

Box 4.2. Measuring the performance of a navigation system

Four parameters are used to measure the performance of a navigation system: accuracy, integrity, continuity of service and availability. These quantities are referred to as the required navigation performance (RNP) parameters. They have their origin in aviation and, taken together, define the level of safety required of a navigation system. The concept has been extended from aviation to marine and land transport. The RNP parameters are:

- *Accuracy:* The degree of conformance of an estimated or measured position at a given time to the true one.
- *Integrity:* The trust placed in the correctness of the information supplied by the navigation system. It includes the navigation system's capacity to provide timely warning to users when the system must not be used for navigation/positioning. Specifically, a navigation system is required to deliver a warning (an alarm) of any malfunction (as a result of exceeding a set alarm limit) to users within a given period of time (time-to-alarm) and with a given probability (integrity risk).
- *Continuity:* The ability of the entire system to perform its functions without interruption during an intended period of operation (POP). The risk is the probability that the system will be interrupted and will not provide guidance information for the intended POP. This risk is a measure of the system's unreliability.
- *Availability:* The percentage of time during which the service is available for use, taking into account all outages, whatever their origins. The service is available if accuracy, integrity and continuity requirements are satisfied.

Source: Ochieng and Sauer (2002).

activities of fishing boats in permitted areas). Other areas of application include science (*e.g.* determination of location and extent of pollution areas, studies of tides and sea level), emergency services and the leisure market segment.

GNSS also provide highly accurate time information that is useful in many areas. Wireless telecommunication networks can use it for network management, for time tagging and for synchronisation of the many frequency references; power plants and networks can use the common time reference system for time stamping, but also as a common reference for system monitoring and control. Finally, the combination of GNSS with wireless communications networks creates opportunities for developing a broad range of location-based services.

Currently, GPS is the only fully operational GNSS and provides only one such signal for non-military use.[35] While it is adequate for many applications (*e.g.* route guidance and information systems), it is insufficient when high levels of accuracy and availability are required.[36] Differential positioning techniques and augmentation systems have been developed to eliminate or reduce some of the common mode errors that affect the global system. The techniques involve estimating the errors in the ranges to the satellites observed at a given time and at a known point. These data (differential corrections) are then broadcast either using terrestrial radio links or satellites (*e.g.* geostationary Earth orbiting satellites) to other receivers in the vicinity to determine their position more accurately, using corrected ranges.

Since the advent of the concept, augmentation systems that can operate on a local, regional and wide-area basis have been developed. Examples of differential services include Omnistar, SwiPos, SaPos and LandStar. Other signal augmentation systems are under development which should considerably enhance the availability of the navigation signals. These include space-based augmentation systems (SBAS), such as the US Wide Area Augmentation System (WAAS) and the European Global Navigation Overlay Service (EGNOS).[37] While the deployment of these and other SBAS around the world represents a major step forward, it is still unclear whether this will be enough for important large-scale civil applications, such as precision aircraft navigation. Another limitation of existing systems is that they do not offer guaranteed service. This may restrict private investment in the development of new systems.

The development of the European system Galileo should overcome some of the difficulties, because it will provide more signals and offer guaranteed service. Galileo will begin with five signals: one freely available to all, like the GPS C/A signal; a commercial service which is more precise; a "safety-of-life" service that can be used for critical applications such as automatic landing of aircraft; a public regulated service (PRS) that will be used by EU governments and, presumably, their armed forces; and a fifth, unique service that combines positioning information with a distress beacon, for use by ships at sea or intrepid mountaineers.

If implemented as planned, Galileo offers a number of advantages for civil use:

- It is designed and developed as a non-military application, while incorporating all the necessary protective security features. Unlike GPS, which was designed for military use, Galileo will provide, for some of the services offered, the very high level of continuity required by modern business, in particular with regard to contractual responsibility.
- It is based on the same technology as GPS and will provide a similar – and possibly higher – degree of precision, thanks to the structure of the constellation of satellites and the planned ground-based control and management systems.
- It will be more reliable, as it will include a signal "integrity message" informing the user immediately of any errors. In addition, unlike GPS, it will be possible to receive Galileo in towns and regions located in extreme latitudes.
- It will represent a public service and, as such, will guarantee continuity of service provision for specific applications. GPS signals, on the other hand, have, on several occasions in recent years, become unavailable on a planned or unplanned basis, sometimes without prior warning.

A recent development that should considerably enhance the quality of the space signals available to civil users worldwide is the recent US-EU interoperability agreement (June 2004) on GPS and Galileo. Interoperability is indeed useful:

- Using both infrastructures in a co-ordinated fashion (double sourcing) offers real advantages in terms of precision and security, should one of the two systems become unavailable.
- The existence of two independent systems is of benefit to all users since they will be able to use the same receiver to receive both GPS and Galileo signals.

An important element of the agreement is the US commitment to ensure that the third generation of GPS (GPS III) will conform to the overall agreement. This will aid the interoperability of the two systems, which is a commercial goal of both sides, and provide a general framework for interoperability with potential future systems developed by third countries.

In addition to interoperability with Galileo and other GNSS, the deployment of GPS III satellites should further enhance global navigation capabilities. A key feature of the planned GPS III satellites is on-orbit adaptability via software uploaded from the ground. GPS III satellites should also be far more resistant to enemy jamming attempts than their predecessors. The ability to improve satellites via software uploads would enable the US Air Force to take advantage of new technology as it becomes available, rather than waiting for a new generation of spacecraft to be developed and deployed. Among the GPS III upgrades possible via software uploaded from the ground are signal power increases and changes to the configuration of the satellites to make them compatible with other satellite navigation systems, such as Galileo. Upgrades not envisioned at this point are possible. Current plans for the GPS III system call for launching the first satellite in 2012, but the Pentagon, facing some Congressional pressure to accelerate the schedule, is studying the possibility of moving the first launch date up to 2009, as was originally planned.[38]

The role of private actors

Because GPS and GLONASS were originally developed for military purposes, both were set up as public systems. However, the gradual emergence of civil applications – both public and private – based mostly on GPS has increasingly raised questions regarding the role of private-sector actors. Over the years, such actors have been more and more involved, not only in the development of the GPS system, but also in the development and operation of differential GPS, as well as in the production of terminals for the professional and mass markets.

The "no fee" approach adopted for GPS has stimulated the growth of commercial GPS applications. It was viewed as a technical necessity arising from the nature of the signals. Moreover, once the signal was made available at no charge, there was no turning back, since it became difficult, if not impossible, to enforce payment afterwards. From a US strategic perspective, a free signal policy had another merit: not only did it stimulate the development of commercial GPS applications, it was also thought that it would minimise incentives for competitors to enter, since it is difficult to compete against a free service.

Ironically, the result was the opposite. In addition to strategic signal independence objectives, the commercial potential of navigation products – largely demonstrated by the commercial success of GPS products, spurred by the availability of the free signal – is one of the main rationales for the development of Galileo. Europe wants its "fair share" of this promising market, which, according to some, could create as many as 140 000 jobs in Europe.

The development of GPS-based civil applications has involved a broad range of public and private actors. In addition to the provider of the GPS signal itself, a first set of other actors offering augmented signals has emerged. A second set of essentially private actors produce the hardware: component manufacturers and product equipment manufacturers.[39] Finally, a third set of actors provide the location-based services that take advantage of the space signals. They can be either public or private, depending on the nature of the application.

The range of services to be offered is potentially very wide. There are strong expectations that location-based services will generate substantial economic benefits, notably by supporting more cost-effective road traffic management, as well as new revenues and jobs for different sectors. According to a study by PricewaterhouseCoopers (PWC) for the European Commission, the market for road transport applications of GNSS (including cars, light commercial vehicles, truck and buses) could exceed EUR 100 billion a year by 2015 (Poulter, 2002). This suggests significant interest from the automobile industry as well as the insurance sector and the tourism industry, for instance. Already, in North America alone, sales of GPS equipment in 2003 were estimated at between USD 3.4 billion (Frost and Sullivan, 2003) and USD 4.7 billion, with asset tracking and fleet management accounting for USD 670 million of the total (Bates, 2003).

A large share of the benefits generated by navigation services takes the form of public goods captured by final users (*e.g.* improved safety of air travel, reduced congestion), while most of the commercial revenues are likely to be captured by downstream service providers. Hence, since the signal has both public and private uses, an interesting question from an institutional perspective, when developing a new system such as Galileo, is what institutional arrangement is likely to be the most effective for the development, deployment and operation of the system.

One school of thought would argue that the public model is the most appropriate, given the significance of the public benefits generated by the system and the strategic dimension of the service. It would also be the simplest way of implementing the system, as it would not require the very complex arrangements involved in the creation of a PPP. Developing such a public system would also be feasible financially for governments. For instance, building Galileo is not expected to cost more than the construction of 150 km of highway in Europe, and the public investment would be justifiable from a cost/benefit perspective. The cost/benefit ratio that has been estimated for Galileo, based on consideration of only some of the potential benefits, is indeed higher than for most public investments.

Another school of thought would argue that, given the commercial revenue that the system is expected to generate, its development should be left entirely to the private sector. In such a scheme, the private operator would have three main sources of revenue:

- Payment for the use of the enhanced public signal by the government, which could take the form of a fixed guaranteed level of revenue for such services.
- Royalty payments by the producers of the chipsets.
- Service fees paid by providers of value-added services that make use the commercial signal.[40]

4. FRAMEWORK CONDITIONS: INSTITUTIONAL ASPECTS

While such a solution may look attractive to resource-strapped governments, it is unlikely to attract private investors. In the case of Galileo, for instance, studies conducted by PWC suggest that while the development of the system should generate substantial benefits from an overall societal perspective, a private concessionaire on its own would not be able to generate sufficient commercial revenue to justify investing in such a system. Moreover, as appeared in the third phase of the project, the success of a private model would depend on a number of critical factors – which represent necessary but not sufficient conditions – both for the business model adopted by the operator and the business environment in which the signal would be provided (see Annex A).

The third and last option is to adopt a PPP approach to the problem, in which governments would provide sufficient resources in the development, deployment and operation phases of the system to attract complementary private investment and make the operation of the system by a private concessionaire economically viable. In spite of the complexities of the necessary arrangements, a PPP offers a number of advantages:

- The private concessionaire has a strong incentive to deliver quality services throughout the lifetime of the concession so as to attract as much business as possible, as well as to establish and develop new market-oriented commercial operations that will generate jobs and revenues.
- Risks are allocated to those best placed to manage them. The concessionaire is best equipped to estimate the business risks involved in providing the service and has strong incentives to keep costs down, to ensure that deployment of the system takes place in a timely manner and that the systems delivered are of high quality.
- The commercial operation establishes long-term revenues, reducing the long-term burden on the public purse.

In the case of Galileo, these advantages have proved attractive to both public and private actors. From a public-sector perspective, the PPP approach allows the Europeans to achieve their strategic objective of signal independence while realising substantial socio-economic benefits and limiting public costs. For private actors, the PPP approach provides opportunities to develop new products and services and tap new markets. Nevertheless, the implementation of the Galileo PPP has raised a number of issues in practice. One of the first was competition issues at the development stage.[41] A further complication, from a competition policy perspective, is the fact that most of the main players are part of Galileo Industries, a European company established in 2000 as a joint venture of leading European space companies to act as the industrial prime contractor for developing and delivering the Galileo infrastructure (corporate members of Galileo Industries include EADS Astrium, Alcatel Space and Alenia Spazio). Therefore, does the fact that ESA has contracted with Galileo Industries for the development phase create a conflict of interest to the extent that companies in Galileo Industries are also bidding for the concessionaire contract?[42]

There are also uncertainties regarding both the revenue the concessionaire may reasonably expect from the public and private sectors and the costs of the system. One concern on the revenue side relates to the potential competition from terrestrial technologies, notably the development of GSM and UMTS-based location services. If such technology develops, the lucrative location-based services market could very well disappear for GNSS, leaving the space-based systems in niche markets.[43]

Uncertainties also exist with regard to public revenue. These revenues come from two sources: i) a fee for access to the PRS; ii) a flat service-availability payment that governments would make in return for the Galileo signals. This could be paid for a number of years. The public sector's guaranteed annual fee should drop after a couple of years, once the Galileo operating company approaches positive cash flow. The uncertainties arise from the fact that some European governments are opposed to the use of the PRS by European military. There are also uncertainties on the cost side of the equation. In particular, it appears likely that additional public funding may be required, both from the Commission and from ESA during the deployment phase. While these problems need to be addressed, they have a good chance of being resolved. For instance, competition problems are clearly inevitable in such a highly concentrated industry. This is not a new problem and has not really prevented the cost-effective development of space systems in Europe in the past, when compared to space initiatives in other parts of the world. Moreover, one may argue that, by their involvement in Galileo Industries, bidders for the concession are well placed to assess the costs and technical risks involved – notably in the development and deployment phases – and therefore to make a realistic submission. Moreover, it is expected that a clearer procurement policy will be established in the near future as part of the work undertaken for the development of a European space programme, to be reviewed by the Space Council by the end of 2005.

Good signs that such problems will be resolved include the strong consensus among governments that the project should go forward, as reflected by their decision to increase their contribution to its financing. The other positive sign is the high reported quality of the submissions of the two candidates for the concession and their desire to proceed.[44]

From an institutional point of view, the Galileo PPP represents an interesting and promising experiment in governance, notably when one takes into consideration that it involves the future provision – over several decades – of services that do not yet exist. It provides an innovative model for generating synergies between public and private actors, with a view to creating maximum benefits for society at large. If successful, it may well inspire other similar initiatives, both within and beyond the space sector.

Notes

1. Although a "privileged" position in government certainly helps space agencies in terms of its programmes, in some cases, it does not always mean a large budget (Verger et al., 2002).

2. OMB's predominant mission is to assist the president in overseeing the preparation of the federal budget and to supervise its administration in executive branch agencies. In helping to formulate the president's spending plans, OMB evaluates the effectiveness of agency programmes, policies and procedures, assesses competing funding demands among agencies and sets funding priorities. OMB ensures that agency reports, rules, testimony and proposed legislation are consistent with the president's budget and with Administration policies.

3. Its space budget is difficult to estimate, since space is not identified as a separate line item in the US budget. Some 2004 figures from DoD show a total (classified and unclassified) space budget of USD 19.4 billion for fiscal year (FY) 2003, USD 20 billion for FY 2004, and a FY 2005 request of USD 21.7 billion. DoD space issues include management of programmes to develop new early warning and missile tracking satellites, and management of military and intelligence space activities generally (US CRS, 2004).

4. The SPOT satellite Earth Observation System was designed by the CNES, the French Space Agency, and developed with the participation of Sweden and Belgium.

5. The other organisations are: the former Industrial Research Promotion Fund, the Bureau for International Research and Technology Co-operation and the Technology Impulse Society (TIG).

6. The current member states are: Austria, Belgium, Denmark, Finland, France, Germany, Ireland, Italy, Norway, the Netherlands, Portugal, Spain, Sweden, Switzerland and the United Kingdom. Canada takes part in some projects under a co-operation agreement. As decided at the ESA March 2004 Council meeting, Greece and Luxembourg will become full members in December 2005.

7. Currently, the CNES is a shareholder in 11 companies (*e.g.* Arianespace, Spot Image); a participant in three economic interest groupings (*e.g.* MEDES); and in five public interest groupings (*e.g.* Médias France).

8. The Norwegian Space Centre is a shareholder since 1994 of two limited companies, the Andøya Rocket Range AS and Norsk Romsenter Eiendom AS), and has the authority as a governmental agency since 2004 [to ensure the state's interests in those two limited companies (NRS, 2004).

9. The Commission recommended notably merging promptly a number of disparate space activities, adjusting chains of command, opening lines of communications and modifying policies to achieve greater responsibility and accountability. Finally, it recommended that the government should sustain its investment in enabling and breakthrough technologies in order to maintain its leadership in space.

10. Military space investment has moved to the top of the agenda of the DGA on orders from the French defence ministry. DGA has already agreed to boost its space research budget by 60% from EUR 50 million to EUR 80 million. Much of the work will be co-managed with the French space agency, CNES (de Selding, 2005).

11. Such co-operation is not always easy to achieve in practice. For instance, a 1998 report from the-then US Government Accounting Office (GAO), noted that the promise of closer NASA/DoD co-operation and the development of a national perspective on aerospace test facilities remained largely unfulfilled (GAO, 1998).

12. One of the most important joint collaborative efforts currently under way between NASA and the Air Force is the National Polar-orbiting Operational Environment System (NPOESS) a tri-agency programme of NASA, the DoD, and the Department of Commerce (DoC), that brings together the DoD and DoC/NOAA polar-orbiting weather satellite programmes. NASA, working with the NPOESS Integrated Program Office, is providing pre-operational risk-reduction demonstration and validation tests for four critical NPOESS sensors that will fly on the NPOESS Preparatory Project (NPP). NPP is a primary NASA mission that serves as a "bridge" between the Earth observation satellite mission and NPOESS. It is also a critical risk-reduction mission for the Visual Infrared Imager Radiometer Suite, the Cross-track Infrared Sounder, the Advanced Technology Microwave Sounder and the Ozone Mapper/Profiler Suite sensors and serves as an end-to-end test for the Command, Control and Communication (C3) and data processing systems for NPOESS (US House of Representatives, Subcommittee on Space and Aeronautics, 2004).

13. Such co-operative successes serve as good arguments for promoters of the creation of dedicated international bodies for space exploration. As seen in Chapter 1, one scenario even includes the possible establishment of an International Space Agency (ISA), following the European Space Agency model, to facilitate such efforts.

14. The main rule adopted by the Agency since its Council at Ministerial level in March 1997 is that the ratio between the share of a country in the weighted value of contracts, and its share in the contribution paid to the Agency, must be a certain percentage (90%) by the end of a given period. That ratio is called the industrial return coefficient. The achievement of geographical return is monitored globally over determined periods. Nevertheless, particular geographical return constraints may be imposed on optional programmes and on mandatory activities, to ensure that they cannot contribute to an unbalancing of the overall return situation (see www.esa.int/home-ind/index.html).

15. ESA has an annual budget of around EUR 3 billion, but this constitutes only half of the annual expenditure in Europe of EUR 5.5-6 billion on civil space activities. The rest comes from investment by national agencies such as CNES, ASI, DLR, etc., in addition to their ESA contributions. The European Commission was to spend about EUR 250 million on space research in its 6th Framework Programme for Research and Technological Development (EC, 2004).

16. Two factors are at play here. The first is that the partners do not have the same legal obligations under the IGA. For the United States, the IGA falls into the category of an "executive agreement", meaning that it does not have precedent over national laws, hence does not bind the US Congress, which can scale down the US contribution. For most European countries, the IGA has instead the status of a binding international treaty. A second factor concerns the differences in the partners' funding procedures: in most European countries, internal procedures ensure funding for agreed programmes throughout several budgetary periods until the programme has been completed, the only reservation being a ceiling on expenditure. ESA uses a multi-year funding procedure with member states' commitments normally covering a three-year period. NASA, on the other hand, must defend annually its programmes and their funding level in Congress.

17. The report was issued in June 2004 by a Commission (chaired by P. Aldrige), created by the President G. Bush, to make recommendations regarding the president's January 2004 "Vision for Space".

4. FRAMEWORK CONDITIONS: INSTITUTIONAL ASPECTS

18. The JSF is a multi-billion dollar US project to replace ageing fighter aircraft on both sides of the Atlantic. The structure of this international programme is based on a complex set of relationships involving both government and industry from the United States and allies that can become JSF partners at one of three participation levels based on their financial contribution, but a predetermined level of work based solely on their financial contribution is not guaranteed.

19. The CGMS came into being in 1972 when several meteorological organisations met to discuss the possible compatibility of their geostationary satellites to cover the whole Earth and reviewed important technical and operational issues (*e.g.* transmission of Weather Facsimile to users worldwide). It became an informal international group, meeting annually, that co-ordinates the independent national or regional systems that form the global system of operational meteorological satellites. The membership includes the operators of meteorological satellites: China, EUMETSAT (for Europe), India, Japan, Russia, the United States and the WMO in its capacity as a major user organisation.

20. The Charter was signed on 20 October 2000. It includes ESA, CNES, CSA, NOAA, ISRO, CONAE, soon JAXA, and potentially other future systems developers. Through the Charter, an authorised user can call a single number to request the mobilisation of the space and associated ground resources of the Charter members in order to obtain data and information on a disaster, free of charge or for a very low fee, usually the costs of reproduction. The "best effort" principle is one of the system's key aspects, as satellites are already often tasked to do other observation missions.

21. The Indian Space Research Organisation established Antrix Corporated Ltd. in 1992 for the promotion and commercial exploration of products and services from the Indian Space Programme. The company's activities are predominantly export-oriented and market in particular software and small spacecraft components (*e.g.* propulsion system valves). Its financial performance has been quite good, with growth of 30% in 2003 and sales turnover of around USD 40 million (Morring and Neelam, 2004).

22. The partial commercial model applies as well to "public" television networks that derive part of their revenues from the sale of commercials.

23. For instance, the penalty imposed on Boeing by the DoD in the aftermath of the Darleen Druyun affair (the former No. 2 acquisition official at the US Air Force was sentenced to nine months in prison for discussing a job with Boeing while overseeing a USD 23 billion lease-purchase deal for the company to supply the Air Force with refuelling tankers) penalised not only Boeing but also the DoD which was forced to drastically reschedule its launch programme and use more expensive alternative launchers.

24. Created in 1979, Inmarsat was a maritime-focused intergovernmental organisation, whose purpose was to improve maritime communications and radiodetermination capabilities of ships at sea (in particular ensuring distress and safety of life at sea communications).

25. Agora stands for *Accès garanti et optimisé pour les régions et l'aménagement du territoire* (affordable and guaranteed offer for rural access).

26. The Skynet 5 model has attracted the attention of other governments. For instance, the French DGA has indicated an interest in moving to a model whereby the government purchases capacity from industry instead of acquiring satellites (de Selding, 2005).

27. WildBlue is scheduled to start broadband service in 2005, targeting US households and small companies in rural areas (the SOHO market) that may never have access to terrestrial broadband services. A second satellite, WildBlue-1, is being built by Loral Space Systems. Its launch is to be determined by market demand.

28. EUMETSAT lists its member states as: Austria, Belgium, Denmark, Finland, France, Germany, Greece, Ireland, Italy, Luxembourg, the Netherlands, Norway, Portugal, Spain, Sweden, Switzerland, Turkey, the United Kingdom. It has nine co-operating states (Slovak Republic, Hungary, Poland, Croatia, Republic of Serbia and Montenegro, Slovenia, Romania, the Czech Republic and Latvia).

29. Three providers of commercial imagery currently participate in the programme: Space Imaging, Digital Globe and Orbimage.

30. About 60% of Spot Image's revenue of EUR 51 million (USD 59 million) in 2002 came from military users, either directly or through Spot partners, in Europe, the United States, South America and Asia. Spot Image's two biggest export markets are the United States and China, with both markets dominated by government demand.

31. NextView is a follow-on programme to ClearView. Launched by the NGA in September 2003, it is intended, in part, to spur the development of a new generation of commercial imaging satellites.

32. Space Imaging, which was recently passed over for a USD 500 million satellite imagery contract with the US National Geospatial Intelligence Agency, faces an uncertain future. Without the NGA's NextView contract, the company does not have the funding to build the two new imaging spacecraft it proposed in its bid.

33. The first satellite navigation system was Transit, deployed by the US military in the 1960s and used extensively by the Navy. It was followed by the development of the more advanced GPS, which offered higher accuracy and stable atomic clocks. The first GPS satellite was launched in 1978. The original use of GPS was essentially military as a "force multiplier", i.e. to provide positioning, navigation and weapons-aiming systems to replace not only Transit, but other US ground-based navigation systems as well. In Russia, the GLONASS system is comparable to the American GPS system, but it is not fully operational mainly owing to financial difficulties.

34. However, such capability was limited by the artificial degradation of the signal through the process of "selective availability" (SA). With SA operational, the instantaneous horizontal positioning accuracy was only 100 meters 95% of the time. In response and in the context of increased civil and commercial utilisation of navigation, terrestrial differential GPS (DGPS) systems were developed to allow greater accuracy of the civil signal. On 29 March 1996, a Presidential Decision Directive was signed by President Clinton that described GPS as an international information utility. As a consequence and due to strong lobbying of civil agencies (e.g. FAA), the Selective Availability was turned off in 2000, as President Clinton announced that the US government will no longer scramble signals from the GPS satellites.

35. GPS satellites transmit two different signals: the precision or P-code and the coarse acquisition or C/A code. The P-code provides what is called the Precise Positioning Service (PPS), a highly accurate positioning, velocity and timing service available only for governmental organisations such as the US military. The C/A-code is designed for use by non-military users and provide the Standard Positioning Service (SPS) for worldwide civil use, but real-time integrity and availability are not guaranteed. The C/A-code is less accurate and easier to jam than the P-code. It is also easier to acquire, so military receivers first track the C/A-code and then transfer

4. FRAMEWORK CONDITIONS: INSTITUTIONAL ASPECTS

to the P-code. The US military can degrade the accuracy of the C/A-code by implementing a technique called selected availability. By 2005, GPS should provide civil users with an enhanced differential frequency capability as the GPS IIR and GPS IIF systems become operational.

36. The limitations of GPS for civil use have been tested, for instance, by Ochieng and Sauer (2002) in the case of advanced transport telematics (ATT) systems. The authors note that GPS suffers from signal blockage (from objects such as buildings, tunnels, trees, canyons, large vehicles) and that, although the system was originally designed to withstand interference and jamming, experience has shown that it is susceptible to these phenomena (Wohlfiel and Tanju, 1999). These are major problems for a number of applications, notably in urban areas. On the basis of an experiment conducted in central London, the authors note further that in the post-selective availability environment, the accuracy required by most advanced transport telematics systems' navigation requirements can be achieved by stand-alone GPS navigation without the need for differential positioning. However, the availability of the required accuracy is very low and is even lower if fault isolation (integrity) is required. They conclude that to achieve the RNP for ATT services, augmentation of GPS is required.

37. WAAS, built by Raytheon Co., based in Lexington, Mass., consists of a ground-based network of GPS receivers and leased transponders aboard a pair of Inmarsat satellites. Lockheed Martin received a contract from the FAA in March 2003 to procure transponder capacity on a third geostationary satellite to broadcast WAAS signals. The third transponder, which will be specifically designed for the WAAS system, is scheduled to be in place by 2006.

EGNOS is Europe's first venture into satellite navigation. It will augment the US GPS and Russian GLONASS systems, and make them suitable for safety-critical applications such as flying aircraft or navigating ships through narrow channels. It will eventually be integrated in the Galileo navigation system. Consisting of three geostationary satellites and a network of ground stations, EGNOS will achieve its aim by transmitting a signal containing information on the reliability and accuracy of the positioning signals sent out by GPS and GLONASS. It will allow users in Europe and beyond to determine their position to within 5 m compared with about 20 m at present (ESA, 2004b).

Another interesting example of a space-based augmented system is provided by the Japanese Quasi Zenith Satellite System (QZSS), or Jun-Ten-Cho in Japanese. The system is developed by the Advanced Space Business Corporation team, including Mitsubishi Electric Corp., Hitachi Ltd., and GNSS Technologies Inc. It is likely to cost around JPY 100 billion and is scheduled to start launch towards the end of the decade into an orbit optimised to cover the Japanese territory. QZSS would provide a new integrated service for mobile applications in Japan based on communications – video, audio, and data broadcasts – and positioning. Its positioning capabilities would, in effect, represent a new-generation GPS space augmentation system, with limited navigation capabilities. In other words, although the QZSS is seen primarily as an augmentation to GPS, without requirements or plans for it to work in standalone mode, QZSS can nevertheless provide limited accuracy positioning on its own. The service also can be augmented with geostationary satellites in Japan's MTSAT Satellite-based Augmentation System (MSAS) currently under development, which features a geostationary satellite-based design similar to the US Federal Aviation Administration's WAAS (Petrovski, 2003). Other systems are currently developed nationally, such as India's GPS and GEO Augmented Navigation programme (GAGAN), or the Canadian Wide Area Augmentation System (CWAAS).

38. The high-level interoperability agreement between the EU and the United States regarding GPS and Galileo is encouraging greater transatlantic industrial co-operation. For instance, Boeing awarded in January 2005 a study contract to Alcatel as part of the US Air Force's GPS III System Preliminary Definition phase. This is the first contractual link between EU and US industry leaders in satellite navigation following the EU-US high-level agreement signed in June 2004. Boeing is also a member of the two consortia (iNavsat and Eurely) competing to become the concessionaire of Galileo.

39. The component segment is mostly composed of satellite navigation chipset producers (*e.g.* semiconductor manufacturers), that may also be present in the production of navigation equipment. Equipment manufacturers includes different types of interrelated actors such as: producers of consumer electronics that integrate the chipsets into receivers and value-added products (*e.g.* Garmin, Magellan) and digital content providers, including producers of digital maps for vehicle, Internet/wireless and business applications (*e.g.* Navteq).

40. For instance, in the business model developed by PWC for Galileo, it is estimated that the royalties on chipsets should represent the principal source of revenue in the early years of the system, while service revenues should develop over time as the market evolves (Poulter, 2002).

41. On the one hand, the Commission would like to see all contracts for the development of the system issued on a competitive basis. On the other, the agreement reached within ESA calls for the application of the geographical return principle to most of the investment (Galileo participants will see 90% of their total investment returned to their countries through contracts to their domestic industry).

42. A private-sector concession, the Galileo Operating Company (GOC) should take over the deployment of the bulk of the 30-satellite constellation at the end of the development phase in 2006. The GOC is expected to finance EUR 1.5 billion of the total cost of Galileo, estimated at EUR 3.5 billion. Galileo operations and maintenance, including as-needed satellite replacement, are estimated to cost EUR 220 million a year. A major task assigned to the current Galileo Joint Undertaking is to organise tenders for the selection of the concessionaire. A first selection was completed in March 2004. An invitation to tender was launched for the second phase in early summer 2004. As of November 2004, only two candidate concessionaires remain in the run: the iNavsat consortium led by EADS Space, Thales and Inmarsat; and the Eurely consortium led by Alcatel Space and Finmeccanica. The final bids were submitted on 25 January 2005.

43. According to Eurely (2004), one of the two consortiums bidding for the Galileo concession, the largest Galileo markets will be for mobile telecommunication location-based services (30% of revenues), transport applications (maritime: 10%, aviation: 13%, road: 14%, rail: 5%), and the use of the PRS by public authorities (20% of the total). On a per signal basis, revenues are expected to be shared as follows: 47% for the open and commercial signals, 31% for the safety-of-life signal, 20% for the PRS.

44. According to ESA officials, both bids are of equal merit and it appears that of the around EUR 2.1 billion that Galileo's deployment and operations phase is expected to cost, two-thirds can be financed by the private sector.

Bibliography

Aldrige, P. (2004), *Report of the President's Commission on Implementation of United States Space Exploration Policy: A Journey to Inspire, Innovate and Discover*, June.

Bates, J. (2003), "Security Concerns Boosting Sales of GPS Devices", *Space News*, 30 June.

Bates, J. (2004a), "For Imaging Firms, Commercial Business Still Seen as Crucial to Long-term Survival", *Space News*, 18 October.

Bates, J. (2004b), "Changing Fortunes at Orbimage", *Space News*, 8 November.

Brachet, G. (2004), "International Cooperation in Earth Observation from Space", paper presented at the Space Policy Institute symposium on "Space Exploration and International Cooperation", Washington, DC, 21-22 June.

CNES – Centre National d'Études Spatiales (2004), *The Agora Initiative*, Release 2.0, CNES Délégation aux programmes radiocommunications, 17 August.

Conseil Économique et Social (2004), "La politique spatiale de recherche et de développement industriel, avis et rapports du Conseil Économique et Social", Report presented by Alain Pompidou, 23 June.

EC – European Commission (2003), *White Paper – Space: A New European Frontier for an Expanding Union: An Action Plan for Implementing the European Space Policy*, European Commission, Brussels.

EC (2004), "Inception Study to Support the Development of a Business Plan for the GALILEO Programme: Executive Summary Phase II", Brussels.

ESA – European Space Agency (2004), European Space Agency's Navigation Web site: *www.esa.int/export/esaNA/GGG63950NDC_index_0.html*, accessed October 2004.

EUMETSAT – European Organisation for the Exploitation of Meteorological Satellites (2003), *Annual Report*, Darmstadt, Germany.

Eurely (2004), European Satellite Navigation Co-operation Day, Warsaw, 30 November.

GAO – General Accounting Office (1998), "Aerospace Testing: Promise of Closer NASA/ DOD Cooperation Remains Largely Unfulfilled", NSIAD-98-52, 3 November.

Goldin, Daniel S. (2004), "Bold Missions and the Big Picture: The Societal Impact of America's Space Program", *Technology in Society*, Vol. 26, No. 2/3.

Morring, F. and M. Neelam (2004), "Application Driven: From the Beginning, India's Satellites Have Served Its Villages", *Aviation Week and Space Technology*, 22 November.

NRS – Norsk Romsenter – Norwegian Space Centre (2004), *Annual Report 2003*, NRS-Report (2004)3, Oslo, July.

Ochieng, W.Y. and K. Sauer (2002), "Urban Road Transport Navigation: Performance of the Global Positioning System After Selective Availability", *Transport Research Part C* (2002), pp. 171-187.

Petrovski, Ivan G. (2003), "QZSS – Japan's New Integrated Communication and Positioning Service for Mobile Users", *GPS World*, June.

Poulter, T. (2002), "Galileo – The Commercial Structure and Revenue Opportunity", PriceWaterhouseCoopers.

Pryke, I., L. Cline, P. Finarelli and G. Gibbs (2002), "Structuring Future International Cooperation: Learning from the ISS", Paper presented at the International Space University's Symposium "Beyond the International Space Station: The Future of Human Spaceflight", June.

de Selding, P. (2003), "Spot Image Focuses On Serving Its Government, Military Customer Base", *Space News*, 18 November.

de Selding, P. (2004), "Congress Modifies IPO Requirements for Inmarsat, Intelsat", *Space News*, 18 October.

de Selding, P. (2005), "French Military Agency Makes Space a Top Priority", *Space News*, 31 January.

Taverna, M. and A. Nativa (2004), "Passing the Baton", *Aviation Week and Space Technologies*, 4 October.

United States House of Representatives, Subcommittee on Space and Aeronautics (2004), "NASA-DoD Cooperation in Space and Aeronautics", Presentation to the House Committee on Science by Mr. Robert S. Dickman, Deputy Under Secretary of the Air Force, 18 March.

US CRS – United States Congressional Research Service (2004), "US Space Programs: Civilian, Military, and Commercial", Report prepared by Marcia S. Smith, Resources, Science, and Industry Division, updated 21 October.

Verger, F., I . Sourbès-Verger and R. Ghirardi et al. (2002), *The Cambridge Encyclopedia of Space*, Cambridge University Press; 1st edition, Cambridge.

WMO – World Meteorological Organisation (2004), "Key Issues: Meteorological Satellites and the World Meteorological Organisation Programmes", Position paper for the OECD Commercial Space Project, February.

Wohlfiel, J.E. and B. Tanju (1999), "Location of GPS Interferers", *Proceedings of the Twelfth International Technical Meeting of the Satellite Division of the Institute of Navigation*, Nashville, Tennessee, September.

ISBN 92-64-00832-2
Space 2030
Tackling Society's Challenges
© OECD 2005

Chapter 5

Framework Conditions: Legal, Regulatory and Public Awareness Aspects

The legal and regulatory framework plays a key role in shaping space activities because it determines the rule of the game under which space actors – private ones in particular – operate. Although a number of basic components of the legal framework are now in place (the international space law regime and some legislation at national level), some major gaps remain. Some space-faring countries still do not have space laws or have embryonic legislation that covers only some types of space activities. Moreover, the regulatory framework does not help to create a stable and predictable environment for business. Problems arise in particular regarding the allocation of spectrum and orbital positions, the liberalisation of space markets remains limited, export controls restrict the ability to exploit market opportunities, space debris continues to accumulate and several standardisation questions remain open.

Another difficulty results from the lack of visibility of space activities in the eyes of the general public. The general perception tends to be distorted, as the media tend to focus on sensational successes and failures so that the general population has a poor understanding of the value of space-based services and is not fully supportive of space activities that could generate substantial socio-economic benefits. This state of affairs reduces the ability of decision makers to take appropriate action in a timely manner for the development of space systems. Moreover, few students are inclined to embrace space-related careers in the current morose context, and the critical pool of expertise that has taken decades to develop is at risk of being eroded.

5. FRAMEWORK CONDITIONS: LEGAL, REGULATORY AND PUBLIC AWARENESS ASPECTS

Introduction

The legal and regulatory framework determines the rules of the game according to which space actors operate. Although a number of basic components of the legal framework are now in place, some gaps remain. As a result, existing regimes are currently not very predictable and/or supportive of commercial space activities. Moreover, the regulatory framework is neither fair nor flexible and does not provide for a level playing field. This can stifle competition and discourage innovation and investment in the development of space systems.

This chapter discusses the relevant legal and regulatory aspects, first considering what has been achieved since the beginning of the space age, at national and international levels, and then identifying major gaps that require policy attention. Next, public awareness aspects are addressed, with a focus on how space activities are perceived by the general public and the potential implications for the development of space systems and the attractiveness of space-related careers for students.

Legal aspects

What has been achieved

Since the beginning of the space age, some progress has been made in developing a legal framework for space activities. At international level, a public law regime has been set up within the context of the United Nations (UN), and multilateral and bilateral agreements have been established to govern co-operative efforts among space-faring nations. Moreover, some nations have developed and implemented national space laws.

An international regime exists

A set of five UN treaties and five resolutions on outer space and space activities provides the foundation for the international regime that governs the relations among nations regarding space. This regime also provides the framework for individual nations' implementation of domestic space laws. Its main focus is on preserving the freedom of exploration and use of outer space, as well as peace and international co-operation, and on defining rights and obligations among states (see Box 5.1).

Box 5.1. **United Nations treaties and main resolutions concerning space activities**

Space activities are regulated at international level by an overarching public regime based on five UN treaties and five resolutions on outer space and space activities. The regime focuses on preserving the freedom of exploration and use of outer space, as well as peace and international co-operation, and on providing for rights and obligations essentially of states.

The core of the regime consists of the following five treaties:

- The Treaty on Principles Governing the Activities of States in the Exploration and Use of Outer Space, including the Moon and Other Celestial Bodies (the "Outer Space Treaty"), opened for signature on 27 January 1967, entered into force on 10 October 1967, with 98 ratifications and 27 signatures (as of 1 January 2003).

- The Agreement on the Rescue of Astronauts, the Return of Astronauts and the Return of Objects Launched into Outer Space (the "Rescue Agreement"), opened for signature on 22 April 1968, entered into force on 3 December 1968, with 88 ratifications, 25 signatures, and one acceptance of rights and obligations (as of 1 January 2003).

- The Convention on International Liability for Damage Caused by Space Objects (the "Liability Convention"), opened for signature on 29 March 1972, entered into force on 1 September 1972, with 82 ratifications, 25 signatures, and two acceptances of rights and obligations (as of 1 January 2003).

- The Convention on Registration of Objects Launched into Outer Space (the "Registration Convention"), opened for signature on 14 January 1975, entered into force on 15 September 1976, with 44 ratifications, four signatures, and two acceptances of rights and obligations (as of 1 January 2003).

- The Agreement Governing the Activities of States on the Moon and Other Celestial Bodies (the "Moon Agreement"), opened for signature on 18 December 1979, entered into force on 11 July 1984, ten ratifications and five signatures (as of 1 January 2003).

The treaties are completed by five sets of legal principles adopted by the United Nations General Assembly. These resolutions promote international co-operation in space activities, the dissemination and exchange of information through transnational direct television broadcasting via satellites and remote satellite observation of Earth, and set some general standards regulating the safe use of nuclear power sources necessary for the exploration and use of outer space:

- The Declaration of Legal Principles Governing the Activities of States in the Exploration and Uses of Outer Space (General Assembly resolution 1962 (XVIII) of 13 December 1963).

> Box 5.1. **United Nations treaties and main resolutions concerning space activities** *(cont.)*
>
> - The Principles Governing the Use by States of Artificial Earth Satellites for International Direct Television Broadcasting (resolution 37/92 of 10 December 1982).
> - The Principles Relating to Remote Sensing of the Earth from Outer Space (resolution 41/65 of 3 December 1986).
> - The Principles Relevant to the Use of Nuclear Power Sources in Outer Space (resolution 47/68 of 14 December 1992).
> - The Declaration on International Co-operation in the Exploration and Use of Outer Space for the Benefit and in the Interest of All States, Taking into Particular Account the Needs of Developing Countries (resolution 51/122 of 13 December 1996).
>
> However, at the most fundamental level, the legal regime for space activities remains that of individual sovereign states. As states are sovereign over their own territory, they have ultimate authority to define the legal environment for any of their national activities, including space activities.
>
> *Source:* UN OOSA (2004).

Aside from the UN treaty regime *stricto sensu*, other sources of space law and regulations include bilateral and multilateral agreements (*e.g.* memoranda of understanding) between governments and/or international and regional organisations. Examples of such agreements include the International Space Station (ISS) Inter-governmental Agreement, mentioned in Chapter 4. To date, the international regime has proved flexible and adaptable enough not to impede the development of space-based applications, whether military, civil or commercial. However, some issues need to be resolved to provide a comprehensive environment for the further development of the space sector.

Some national space laws have been implemented

The national level is still the most fundamental one for the legal regime covering space. Given the liability implications of space activities under the international public law regime, it is in the best interests of space-faring nations to implement dedicated national laws that regulate space activities falling under their jurisdiction.[1]

Several space-faring countries have in fact passed basic or complex legislation to deal with their obligations under international law (*e.g.* United States, Australia, United Kingdom), for instance with regard to the licensing of space activities or the registration of space objects.[2]

The development of an embryonic national space law for specific sectors is often the first step. For instance, in Canada, the need to address the role of commercial actors in Earth observation (EO) activities has led to new Canadian legislation to regulate the operation of remote-sensing space systems. If passed, the legislation, which is under consideration in the Canadian House of Commons (January 2005), will permit Canadian companies to own and operate remote sensing satellite systems, while providing the Canadian government with the authority to order priority access or interrupt normal service to protect national security, defence or international relations interests and to observe international obligations (Foreign Affairs Canada, 2004).

Many space-faring countries have taken into account the fact that national space laws represent a major element of the legal and regulatory environment in which private space actors operate. Such laws establish clearly how their national governments interpret international law, making the rules of the game more transparent for private firms. As legal and regulatory uncertainties are reduced, private actors are in a better position to make sound business decisions. However, few countries have so far enacted integrated national space laws.

Issues to be addressed

A number of issues tend to make the overall legal framework for space activities not entirely predictable or supportive of commercial ventures. First, many space-faring countries still do not have national space laws at all, or have laws that cover only some space activities, thereby creating a legal vacuum. Second, the international legal regime is basically a public law regime which deals with rules and obligations that apply to sovereign states. Hence, it is not really suitable for business activities. Third, there are a growing number of general laws, beyond space law *per se*, which have a major bearing on the success or failure of space applications. Finally, the private financing of assets is more difficult in the space sector than in other industries, in part because of the special legal regime that applies to space activities.

Many countries still do not have national space laws

By subscribing to the international legal regime for outer space, governments have accepted a certain number of obligations when conducting space activities. It is their responsibility – and in their best interest – to ensure that they have a national regime that takes account of these obligations. For instance, governments are liable under international space law whenever a space object is launched from their territory. They can mitigate the related risks by developing an appropriate licensing structure that regulates launching activities taking place on their soil.

Moreover, many national regimes for space activities remain fragmented, focusing on a limited number of areas. France has recently undertaken work to assess the need for an integrated national space law. This space-faring country has an active industry and decades-old institutional programmes, but not yet a comprehensive national space law. The current regulatory system has evolved through an accumulation of contractual or administrative arrangements, as new developments were regulated on a case-by-case basis. This *ad hoc* regime tends to be lacking in clarity for new entrants.

Experts studying this issue have concluded that, while the existing regime is acceptable for current programmes, it needs rethinking in the near future, in particular to accommodate the new commercial systems being developed (ministère délégué à la Recherche et aux Nouvelles Technologies, 2002). They recommended implementing a robust licensing system, with appropriate provisions regarding the role to be played by the CNES.[3] As in many other countries, space applications will have to pass through a variety of governmental agencies for approval.

While legislative developments at national level are necessary, it is important to ensure that such efforts are carried out in a harmonised manner across space-faring countries. In this regard, work conducted in the context of Project 2001 Plus is particularly useful, notably the consideration of the various "building blocks" that need to be part of a commercially oriented national regime.[4] They include notably: authorisation aspects of space activities; supervision of space activities; registration of space objects; indemnification regulations; and procedural implementation of all regulations, including any additional regulations required when conducting commercial activities (i.e. insurance and liability, environmental, financing, patent law and other intellectual property rights, export controls, transport law, dispute settlement).

The international legal regime is not well suited to space business

At international level, a number of issues and uncertainties remain. No state or commercial entity has yet called officially into question the founding principles of the international space law regime. Yet, in the decades to come, this regime will face the challenge of adapting to the increasing commercialisation of space, and some of its main principles will come under pressure.

Terms used in the space treaties need to be defined more clearly

Such terms as "space object", "outer space", or "launching state" are not clearly defined in the space treaties. This opens up room for conflicting interpretations by the increasing number of commercial firms involved in space activities. As an example, the Outer Space Treaty does not define precisely "outer space", and no formally accepted legal delimitation of outer space currently exists. Where does airspace end? Where does space begin? This may have implications for the licensing of future launchers, under either air or space law, and the liability of private entrepreneurs when spacecraft travel through different airspaces and in outer space.

A similar problem is encountered for other terms used in the space treaties. For instance, the current concept of "launching state" may be inadequate to cover all space activities (Box 5.2).

> Box 5.2. **Sea Launch and the concept of launching state in international law**
>
> Under international law, the "launching state" is absolutely liable for damage caused by its space object on the surface of the Earth or to an aircraft (Article II of the Liability Convention). It is liable elsewhere (in space) only if the damage is its fault or that of a person for whom it is responsible (Article III of the Liability Convention). Hence, it is very important, for the compensation of victims on the basis of space law, to be able to identify the "launching state" with certainty. According to Article Ic of the Liability Convention, the term "launching state" means: i) a state that launches or procures the launching of a space object; or ii) a state from whose territory or facility a space object is launched.
>
> The application of the definition to Sea Launch raises a number of issues. Sea Launch is an international launch service provider, formed of a consortium of companies from the United States, Norway, Ukraine, Russia and other countries, which was established in 1995 and became operational in 2000. The main issues raised regarding the definition of launching state are as follows:
>
> - *State:* Sea Launch LLC was originally registered in the Cayman Islands and is a consortium of companies from various countries.
> - *Territory:* The launch is not performed from a "territory" but from a platform on the high seas.
> - *Facility:* The platform was originally registered in Liberia.
>
> The question thus arises whether the concept of launching state is still adequate to cover all forms of space activities, including new private ventures such as Sea Launch, or whether it opens the door to a "flag of convenience" approach to space faring. However, in the specific case of Sea Launch, the problem no longer arises since it is now registered by the Federal Aviation Administration (FAA) in the United States and is headquartered in Long Beach, California.

Rules related to the non-appropriation of outer space and the ownership of space objects need to be reviewed

The Outer Space Treaty states that: "Outer space, including the Moon and other celestial bodies, is not subject to national appropriation by claims of sovereignty, by means of use or occupation, or by any other means" (Article II). This means, in effect, that no one can own any portion of space. This non-appropriation principle, and the derived rules concerning the exploitation of natural resources in outer space, may come under discussion in the future in the context of space exploration for profit by private enterprises (Reif, 2002).

Satellites and other space objects remain the property of their legal owners, regardless of their location. This rule has implications in terms of liability and intellectual property. The Registration Convention of 1976 also requires states to maintain an appropriate registry of the space objects they launch into outer space, and, further, to transmit to the United Nations certain information concerning each space object entered in their registries. Besides providing an up-to-date record of objects in orbit, an international registry could be used increasingly by external private investors to obtain information for assessing loan applications made by satellite operators. However, many countries still do not possess a national public registry that can be consulted by any interested party.

In resolving these issues, legal experts have stressed that efforts should be made to leave the current space treaties untouched, given the traditional reluctance of states to create new binding international instruments.[5] Rather than introducing amendments to treaties in light of changes in the space sector, it is often suggested that separate instruments should be adopted, where necessary, to give a more precise meaning to certain aspects of the treaties or to treat specific items. They could take the form of principles and guidelines, codes of conduct or UN General Assembly resolutions. One example is space debris, which is currently under consideration at the UN Committee on the Peaceful Uses of Outer Space, and could lead in time to an international regime. One exception concerns the Moon Convention, which many would like to see reviewed in order to clarify the conditions of its potential commercial exploitation (ILA, 2002).

The financing of space assets raises a number of issues

In most business activities, it is essential to be able to finance the acquisition of productive assets by borrowing from private lenders. Typically, the productive asset is used as collateral so as to protect the lender against default by the borrower. In the space sector, the range and volume of activities conducted by private actors have increased dramatically over the last decade. Such commercial space systems are extremely capital-intensive and can take years to complete. However, there is not yet an established market for commercial financing of private space activities, as exists for most other industrial sectors. The space industry is often left the option of financing projects through the sale of high-interest bonds, which is not always possible or successful.

To address this problem, the International Institute for the Unification of Private Law (UNIDROIT, see Box 5.3) is currently working to develop a specific space protocol to the UNIDROIT *Convention on International Interests in Mobile Equipment*, which would provide clear financing schemes for space assets.[6] The protocol (formally, the Protocol to the Cape Town Convention on Matters Specific to Space Assets) would establish a framework through which states support a system of asset-based and receivables financing. By permitting secured financing for the space sector, the protocol would have considerable potential to enhance the availability of commercial financing for outer space activities and to further the provision of services from space to countries in all regions and at all levels of development.

More specifically, the protocol could provide an international framework for asset-backed loans, whereby lenders will be able to seize the property in case of default on the loan, to liquidate the assets, to get cash for a defaulted loan, even in the case of international projects. The international rules, enabling creditors to reduce their risk by obtaining the asset as collateral, would govern the seizure of hardware and associated rights such as access to the ground facilities and use of the licences needed to operate it.

Box 5.3. **What is UNIDROIT?**

The International Institute for the Unification of Private Law (UNIDROIT) is an independent intergovernmental organisation located in Rome. It was set up in 1926 as an auxiliary organ of the League of Nations, and, following the demise of the League, it was re-established in 1940 on the basis of a multilateral agreement, the UNIDROIT Statute. Members include 59 states.* UNIDROIT activities are meant to contribute to the development of a reliable and efficient, harmonised international legal framework for public and private actors alike. The institute's main purpose is to study needs and methods for modernising and co-ordinating private and, in particular, commercial law among states and groups of states. Previous achievements have included, for example, the enactment of Principles for International Commercial Contracts (1994, revised in 2004) and a Convention on Stolen or Illegally Exported Cultural Objects (1995).

* UNIDROIT signatory members States include: Argentina, Australia, Austria, Belgium, Bolivia, Brazil, Bulgaria, Canada, Chile, China, Colombia, Croatia, Cuba, Cyprus, Czech Republic, Denmark, Egypt, Estonia, Finland, France, Germany, Greece, Holy See, Hungary, India, Iran, Iraq, Ireland, Israel, Italy, Japan, Luxembourg, Malta, Mexico, the Netherlands, Nicaragua, Nigeria, Norway, Pakistan, Paraguay, Poland, Portugal, Korea, Romania, Russian Federation, San Marino, Serbia and Montenegro, Slovakia, Slovenia, South Africa, Spain, Sweden, Switzerland, Tunisia, Turkey, United Kingdom, United States, Uruguay, Venezuela.

Source: UNIDROIT (2004a).

The protocol would also set up a more efficient international registry for space assets, possibly through the United Nations, to give lenders a place to examine a company's assets and the possible existence of other loans that may have rights to the same hardware in case of default. By reducing the lenders' financial risks, the protocol – if widely adopted – would make it easier for space companies to get loans, and at lower interest rates, thereby lowering their total project costs.[7]

This protocol is well past the preliminary stage, which consisted in the preparation of a first draft, piloted by a Space Working Group. This group, composed of representatives of the manufacturers, financiers and users of space assets, prepared the draft protocol, in liaison with other bodies such as the UN Committee on the Peaceful Uses of Outer Space.[8] The draft is being reviewed by the representatives of governments designated to take part in the Committee of Governmental Experts, which is ultimately responsible for determining the shape of the protocol. The committee has already held two sessions, one in December 2003 and another in October 2004. The next session is due to be held in June/July 2005. A final successful Committee session in 2006 could make it possible to convene a Diplomatic Conference, probably in the first half of 2007.

Governments will need to engage individually in this new treaty regime by signing and then ratifying the protocol (to make it applicable in their national laws), while the investment community will need to be made aware of the new regime.

General (non-space) laws are not always "space-friendly"

As the studies carried out in the third phase of the project (which focused on the development of business models for particular applications) clearly demonstrated, the success or failure of space applications will depend not only on space law *per se* and the way it is applied, but more importantly for most space applications, on general legal provisions far beyond the scope of space law. This is because for the large majority of space applications, the space segment represents a small – albeit essential – component of the value chain. Hence, the laws and regulations that affect other segments of the chain and the final product or services will have at least as much, if not more, of an incidence on the economic feasibility of a particular space application than space law *per se* (Von der Dunk, 1999).

Although the third phase of the project considered quite a broad range of applications, there was a good deal of communality regarding the legal factors that played a particularly important role in the success or failure of the applications. These factors concern in particular competition policy, liability and intellectual property.

Competition

Owing to the international commercialisation of space-related products and services, competition issues arise across many applications. One key issue relates to the need to maintain a level playing field between competing terrestrial and space-based solutions in a commercial market. Important factors to be considered in this regard are: *i)* the extent to which governments have invested in the application and are involved in its operation; *ii)* the role of the regulations that shape the economic environment and whether such regulations are truly technology-neutral; *iii)* the support provided by governments to specific and sometimes competing technologies.

Liability

Liability issues arise in the production and operational phases of the space systems used for running the applications, as well as in the operation of the applications themselves. For instance, application-specific liability issues arise in the field of telehealth (*e.g.* the liability of the health professional conducting a tele-consultation) in addition to the liability faced by the developer and operator of the telehealth network and the operator of the space-based infrastructure used by the network. The same problem arises for location-based services (*e.g.* the liability of the provider of safety-of-life and certified signals) and space tourism (*e.g.* the liability of space entrepreneurs and public authorities under the liability convention for damage to third parties, and their liability *vis-à-vis* their clients). In some cases, special legislation can be introduced to encourage the development of a new application by restricting somewhat the liability risk faced by space entrepreneurs. This is the case, for instance, in the United States for the Commercial Space Launch Act of 1984 and the Commercial Space Launch Amendments Act of 2004 (H.R. 5382) which was passed into law in December 2004 (Box 5.4).

Intellectual property

Intellectual property issues are likely to become more prominent as the private sector plays an increasingly important role in the development of space assets. As current discussions surrounding this issue in relation to the European navigation system Galileo demonstrate, the proper, transparent and effective protection of intellectual property generated in the context of space activities is one of the most important legal issues for private entities interested in investing in space.

As the World Intellectual Property Organisation (WIPO) has noted, the effective acquisition and protection of intellectual property rights would have a positive effect on the participation of the private sector in the development

Box 5.4. **Limiting the liability of commercial launch operators**

A state is deemed liable for compensation for damage caused by its space objects on the Earth, in the air or in space under the Outer Space Treaty (1967, Article 7) and the Liability Convention (1972). Given this, and as commercial actors have been increasingly involved in space activities, a number of space-faring countries have set up specific indemnification regimes to limit the liability of their commercial launch operators in case of accident. With regard to liability of the nascent suborbital activities of commercial entrepreneurs, the United States has established the first specific licensing regime for this type of actor.

Governmental indemnification regime: The US example

- *Objectives:* The US liability indemnification regime is meant to provide US industry and its insurers with some level of financial comfort in the case of a launch mishap, to keep the price of insurance at an affordable level for the industry (without government indemnification, the launch industry would have to spend a significant amount more on insurance), and to keep the US space transport industry on an even playing field with foreign competitors that benefit from equivalent indemnification regimes.

- *Description:* The US indemnification regime limits the liability of US launch operators licensed by the FAA in case of accident. Under the current law, the US government requires that launch operators purchase insurance to cover the first USD 500 million of any third-party claims due to a launch accident, and the government, if necessary, will cover further damages and claims up to USD 1.5 billion. To date, the indemnification provisions have not been exercised; nonetheless, the US government continues to bear a financial risk.

- *Prospects:* Originally, the regime was set to expire five years after its creation in the original Commercial Space Launch Act (CSLA) of 1984, but it has been regularly extended by the US Congress. With the maturation of the commercial space transport industry, the liability indemnification regime may eventually be eliminated or phased out. The recent amendments to the Commercial Space Launch Act of 2004 require in particular a study by the National Academy of Public Administration on how best to gradually eliminate the liability risk-sharing regime by 2008 or as soon as possible thereafter.

The Commercial Space Launch Amendments Act of 2004 (H.R. 5382)

The recent amendments to the Act provide a specific regime for commercial developers of reusable suborbital rockets to experiment and start generating revenue by taking on paying passengers flying at their own risk

> Box 5.4. **Limiting the liability of commercial launch operators** (cont.)
>
> (i.e. sheltering launch companies from second-party – passenger – liability). The bill allows the Associate Administrator for Commercial Space Transportation (AST) in the FAA to issue experimental permits for an unlimited number of flights for a particular vehicle design. The FAA will have to work closely with applicants on a case-by-case basis to determine what modifications may be made to a suborbital rocket without changing the vehicle design to an extent that would invalidate a permit (i.e. legal definitions for "suborbital rocket" and "suborbital trajectory" clarify in the bill the parameters of AST's regulatory jurisdiction). AST's decisions in this regard should be driven by the dual goals of promoting the industry and protecting the safety and health of the general public. The regime is to be amended in eight years, although a provision requires the FAA to take action and set up new regulations before this date in case of any suborbital event that poses a high risk of death or injury.

of outer space activities and on the further development of space technology in general (WIPO, 2004). Moreover, as many space projects involve international co-operation, there is a need for a simple, uniform and reliable international legal framework.

This is not the case today, as nations are unwilling to relinquish their right to issue patents, trademarks, copyrights, etc., to other nations or to international organisations. Although national intellectual property laws are relatively well harmonised, different national laws still apply different principles. Once a dispute arises, national laws regulate questions of international jurisdiction. Thus, a lack of a reliable international legal regime requires parties to negotiate intellectual property clauses in each international co-operation agreement, which may include, for example, issues of ownership, rights of use, rights of distribution and licensing of data, information capable of legal protection and confidentiality. Obviously, while such a contractual agreement is valid among the parties concerned, it does not bind third parties.

As clearly stated by WIPO, "[T]he importance of establishing a legal regime that effectively protects intellectual property in space cannot be overemphasised. Lack of legal certainty will influence the advancement of space research and international co-operation. Because of the large investments involved in space activities, a legal framework that assures a fair and competitive environment is necessary to encourage the private sector's participation. Limited exclusive rights conferred by intellectual property protection would bring competitive benefits to rights holders, either through a licensing agreement or by exclusion of competitors from use of a given technology. Intellectual property rights created in the company may improve the company's overall image. For example, the acquisition of patents may be viewed as a proof of a company's technical competence. The licensing of intellectual property also offers the advantage of being able to negotiate cross-licensing with other parties, particularly when the development of space applications requires the incorporation of various high technologies. Further, legal mechanisms to establish and maintain security interests in intellectual property exist in certain countries" (WIPO, 2004, p. 5).

As general laws beyond the scope of space law often have a major bearing on the success or failure of space applications, it is the responsibility of governments to review such laws to ascertain whether they need to be amended or whether the way in which they are applied to space-related activities should be modified.

Regulatory aspects

What has been achieved

In addition to the legal framework discussed above, the regulations that govern space activities on an ongoing basis will play a central role in shaping the evolution of the sector. Ideally, such regulations should help to create a stable and predictable business environment, while at the same time stimulating innovation and encouraging entrepreneurship.

Some progress in this direction has been achieved over the years, at both national and international levels. For instance, rules have been established at international level for the allocation of geostationary positions and radio spectrum frequencies, while regulations for space telecommunications have been introduced at the national level in several countries. Moreover, the World Trade Organisation (WTO) has liberalised some space telecommunications' services markets. Finally, some efforts have been made to establish technical standards that help to facilitate the development and use of space assets.

Geostationary positions and radio spectrum are allocated internationally

Radio frequencies and the geostationary satellite orbit are limited natural resources and must be used rationally, efficiently and economically. No two radio systems, including satellite responders, can operate on exactly the same frequency and in the same orbital position without causing harmful interference to one another. The International Telecommunications Union (ITU) is managing the global co-ordination of radio frequency applications, largely successfully (Box 5.5).

> Box 5.5. **The allocation of satellite orbital positions and frequency use**
>
> Regardless of whether a new satellite system is to be operated by government or privately, all requests for satellite orbital positions and frequency use are submitted by national administrations on behalf of operators to the International Telecommunications Union (ITU), an international organisation within the United Nations System in which governments and the private sector co-ordinate global telecommunications networks and services. The co-ordination sometimes takes years for complex satellite systems.
>
> **Phase 1: Advance publication information (API)**
>
> The first stage, known as advance publication information (API), sees the national administration supply details on the identity of the satellite system, the expected date of entry into use, and orbital and network characteristics (such as the frequencies requested, and transponder and power emission data). At this early stage, ITU simply ensures that the necessary information is complete and publishes it in the International Frequency Information Circular (IFIC), which is distributed to all ITU members. This provides an opportunity for other governments to determine whether the planned system poses any potential threat to existing terrestrial or satellite services (including those already under co-ordination), in terms of orbital position or interference. Administrations have four months to comment on the proposed system.
>
> **Phase 2: Co-ordination**
>
> This second and more complex phase of the process involves formal co-ordination between the proposing administration and all countries reported by the ITU as being affected by the proposed system. At this stage, ITU also verifies that the proposed system conforms to all relevant provisions of the Radio Regulations, the internationally binding treaty governing the use of the radio frequency spectrum. The proposing administration completes a form providing detailed system characteristics, including the characteristics of Earth stations and their proposed locations (Appendix 4 of the Radio Regulations). Administrations then work multilaterally through a series of co-ordination meetings to resolve difficulties, normally by adjusting the technical parameters of the proposed new system to ensure it does not interfere with existing services.
>
> **Phase 3. Notification**
>
> The final phase of the process, notification, sees the ITU perform a final verification that formal co-ordination has been successfully completed and that the system still conforms to the Radio Regulations. If a favourable finding is given, the system is recorded in the Master International Frequency Register (MIFR). If not, the notice is returned to the proposing administration which then has the option of continuing negotiations until a favourable outcome is achieved or no objections or complaints are made about the transmissions within a four-month period.
>
> *Source:* Adapted from ITU (2003).

The WTO has liberalised some space telecommunications' services markets

As seen in previous sections, market access for space products and services is often restricted. For strategic and/or security reasons, public procurement – the largest segment of the market for space goods and services – is often limited to the national or regional level. This tends to stifle competition and result in a misallocation of resources, since firms that should have exited the industry remain artificially active, while firms that could serve the market more efficiently are barred from doing so.

Within the WTO, a global regime for trade and international liberalisation of telecommunications has been developed on the basis of the General Agreement on Tariffs and Trade (GATT) and the General Agreement on Trade in Services (GATS). The WTO's Agreement on Telecommunications Services, under the Fourth Protocol to the GATS, provides that participating states commit to allow foreign satellite communications operators to offer their services on a reciprocal, non-discriminatory basis in their countries.

Another action taken by the WTO, under the TRIPS Agreement (Trade-related Aspects of Intellectual Property Rights) could also contribute to the global harmonisation of intellectual property aspects (Malanczuk, 1999). However, as WIPO has noted (WIPO, 2004), the TRIPS Agreement does not specifically address the question of outer space as such. In addition to the principle of national treatment in Article 3 of the TRIPS Agreement, Article 4 provides that, in principle, any advantage, favour, privilege or immunity granted by a member to the nationals of any other country shall be accorded immediately and unconditionally to the nationals of all other members ("most-favoured-nation treatment").[9]

Further, according to Article 27.1, patents must be available and patent rights enjoyable without discrimination as to the place of invention. Therefore, national law has to ensure that, with respect to inventions created in outer space, patents must be granted and enforceable in the territory in which they apply under the same conditions applicable to inventions created elsewhere (WIPO. 2004, p. 7). Therefore, while originally developed without specific consideration of space activities, the general WTO regime now has considerable impact on satellite communication services and intellectual property rights.

Some standards have been implemented

Historically, standards in the space sector have been based on military standards and set up independently by space agencies or other technical agencies. Industry standardisation groups currently play an increasingly important role in developing space and non-space standards (Box 5.6).

Box 5.6. **International bodies active in the standardisation of space systems**

Aside from national and regional standardisation bodies (*e.g.* in the United States: the American Institute of Aeronautics and Astronautics [AIAA] Standards Executive Council; in Europe: the European Co-operation for Space Standardisation [ECSS]), there are two main international bodies active in the standardisation of space systems: the Consultative Committee for Space Data Systems (CCSDS); and the International Organisation for Standardisation's Technical Committee "TC 20 Aircraft and space vehicles".

The Consultative Committee for Space Data Systems. CCSDS was created in January 1982. It provides an international forum for space agencies and other organisations and companies interested in mutually developing standard data-handling techniques to support space research, including space science and applications. The governmental or quasi-governmental organisations that meet regularly represent 28 countries (founding members, observers), and over 100 commercial associates participate in specific working groups. To date, 300 missions have flown with CCSDS protocols. In addition, private vendors have developed more than 100 CCSDS-compliant mission-support products, ranging from spacecraft platforms and space-qualified hardware components to ground support hardware and software. The primary products of the CCSDS are technical recommendations that guide internal development of compatible standards within each participating space agency. CCSDS activities are believed to significantly enhance the planning and execution of future co-operative space missions. An intrinsic contribution of the CCSDS recommendations is the expected higher degree of interoperability among agencies that observe them. The fundamental operating principle of the CCSDS is consensus. Under an agreement entered into between CCSDS and the International Organisation for Standardization (ISO) in the mid-1990s, CCSDS acts as the principal technical engine of ISO Technical Committee 20 (TC 20)/Subcommittee 13 (SC 13), and most CCSDS recommendations are processed into full ISO standards via this relationship.

The International Organisation for Standardization's "TC 20 Aircraft and space vehicles". The 1980s saw a demand from the telecommunications industry for a set of international non-governmental standards applicable for commercial space products, and ISO, as the world's largest developer of standards, was the logical forum. ISO standards are voluntary and do not require individual countries to change or discard existing specifications. They only require individual countries to exchange up-to-date information to create common definitions of existing and agreed interfaces, in co-ordination with the relevant industry (ISO, 2004). The Technical Committee (TC) 20 focuses on aircraft and space vehicles. Its scope is the standardisation of materials, components and

> Box 5.6. **International bodies active in the standardisation of space systems** *(cont.)*
>
> equipment for construction and operation of aircraft and space vehicles as well as equipment used in the servicing and maintenance of these vehicles. Two of its subcommittees are dealing particularly with space-related standards (TC 20/SC 13: Space data and information transfer systems, and TC 20/SC 14: Space systems and operations).
>
> - Total number of published ISO standards related to the Committee 20 and its subcommittees: 471.
> - Number of published ISO standards under the direct responsibility of the TC 20 Secretariat: 63.
> - Participating countries: 12; observer countries: 22.
> - International organisations in liaison: AECMA-STAN, Committee on Space Research (COSPAR), European Commission, European Organisation for Civil Aviation Equipment (EUROCAE), International Council of Aircraft Owner and Pilot Associations (IAOPA), International Air Transport Association (IATA), International Civil Aviation Organization (ICAO), World Meteorological Organisation (WMO).
>
> *Sources:* CCSDS (*www.ccsds.org*) and ISO (*www.iso.ch*).

Standards have a strategic dimension, since once they are embodied in national or regional regulations and in industry practices, they can be crafted to either facilitate or impede market access, and they have extensive ramifications for a company's product testing and design. It has therefore become crucial for competitive companies to participate actively in the development of space standards. The main difficulty is the need to get involved in the appropriate arena.

The need to standardise space systems in different countries is relatively recent. It has been spurred by the multiplication of international co-operation projects and by the increasing commercialisation of space-related products and services, satellite communications in particular, so that rather than develop standards themselves, many countries adopt standards issued by various international industry and intergovernmental associations. In Europe, for instance, it was only in the early 1990s that European space-faring countries agreed on the development of common standards for space-related equipment. Although the national agencies' requirements were essentially similar, the impact of the differences in standards led to higher costs, a less competitive industry as a whole and potential errors in co-operative ventures. Therefore, looking beyond product assurance, the European space community agreed in 1993 to create and promote the European Co-operation for Space Standardisation (ECSS) initiative as the central European structure for space standardisation to be used by ESA, national space agencies and industry.

Issues to be addressed

While an overall regulatory framework exists and has allowed and sometimes even fostered the development of space activities, many issues remain unresolved, especially in light of the increasing role of the private sector in space. They include issues related to the allocation of orbit slots and spectrum frequency and the scope for extending WTO discipline to a broader range of space goods and services. Moreover, space debris issues will increasingly need to be tackled, and the standardisation of some space products and services remains a challenge.

The ITU process raises a number of issues

The ITU plays a very important role in co-ordinating spectrum allocations and orbital positions, but huge worldwide demand for satellite-based services has brought out some inherent difficulties related to co-ordination. For instance, the number of satellites providing different types of info-communications services has shown steady growth over the last decades, from 24 in 1985 to an estimated 150 in 2002 (ITU, 2003). As a result, the allocation of spectrum and orbital slots by the ITU has become more complex.

One issue relates to over-filing – often known more familiarly as "the paper satellite problem" (satellites that were never meant to be developed and launched). The "first come first served" rule used by the ITU to process applications gives opportunistic actors an incentive to submit files for satellites, even when they have little hope of actually launching them. Indeed, the past decades have seen many routine requests for orbital positions and frequencies that are not actually needed, with a view to "reserving" those positions and frequency bands for possible future use, or for commercial resale to another user at a later date. Although efforts have been made to address the problem and although the number of "paper satellites" has declined, the current arrangement still only obliges operators to provide details once a system is about to be launched, so that there is no serious dissuasive effect.

Faced by a large and growing backlog of co-ordination requests, the ITU has started tackling the paper satellite phenomenon, which also slows the development of legitimate commercial systems. It has done so through its regular Plenipotentiary Conferences (the supreme organ of the ITU convened every four years), by introducing new "due diligence" administrative and financial procedures to discourage abusive filings, but many entities still do not seriously intend to deploy satellite systems and contribute to the backlog that blocks the overall system.[10]

With regard to the allocation of spectrum, the ITU is also conducting work to try and resolve some of the problems in the framework of a specific resolution on "Options to improve the international spectrum regulatory framework" (i.e. Resolution 951 [WRC-03]).[11] The results of these studies will be included in the Radiocommunication Bureau Director's report to the next world radiocommunication conference in 2007.

Some experts advocate a more radical reform of the frequency allocation process. They argue that the "first come first served" rule currently used by the ITU should be abolished and that future satellite communications spectrum should be auctioned to the highest bidder. This would have the benefit of transparency and would allow more efficient access to existing slots. However, it might also have a major impact on the viability of existing satellite operators. Moreover, if applied across the board, it might have a detrimental effect on developing countries, which are already concerned that, under the current "first come first served" regime, too few frequencies and orbital slots will be available to meet their future needs by the time they are ready to invest in satellites. These concerns could be met in an auction-based allocation system by setting aside slots and frequencies for developing countries, with transparent rules about their use. Alternatively, any money raised through the auction process could be channelled to a fund designed to foster space applications for health, education or environmental purposes in the developing world.

The extent of WTO discipline remains limited

Despite its important role in the liberalisation of telecommunications, the WTO's general role in the area of space activities is still relatively limited. As the importance of commercial activities increases, the costs imposed by trade restrictions (e.g. export controls) are likely to increase and may become counter-productive, not only from an economic perspective, but also perhaps from a strategic one.

In fact, liberalisation of the telecommunications sector – the most advanced commercial space sector – is still incomplete, as restrictive policies remain in force in many countries. Such policies include, in particular, cumbersome administrative requirements, restrictive regulatory procedures and unfavourable treatment for non-national satellite operators. This encompasses discriminatory licensing conditions, restrictive operating conditions, disparate tax or fiscal obligations, requirements to establish a local presence or legal entity, and sometimes use of a predefined business model (ESOA, 2003).

In many countries that are parties to the WTO Agreement on Telecommunications, the main problem lies with the implementation of liberalisation, more than with the relevant legislation. Satellite operators tend to have concerns regarding the future allocation of frequencies, exclusivity of bands and licensing for ground equipment that countries may impose when liberalising spectrum use. This includes additional charges operators may have to pay (*e.g.* for the downlink providing direct-to-home services).

Aside from the specific case of the telecommunications market, many other space-related activities are not yet considered at the WTO. Because many space components and systems are considered as munitions or other military equipment, they are excluded from the mandate of the WTO, which does not intervene in the trade of military systems.

Even so, as commercialisation expands, some efforts might be made in the context of the WTO to liberalise the procurement of space systems by governments. For instance, the Agreement on Government Procurement (AGP) of 1979 has started opening up the business of government procurement to international competition. It is designed to make laws, regulations, procedures and practices regarding government procurement more transparent and to ensure that they do not unduly protect domestic products or suppliers, or discriminate against foreign products or suppliers. As governments represent the largest market for the world space industry, with over USD 42 billion allocated worldwide in 2003 for civilian and military space budgets (Euroconsult, 2004), opening up government procurement when possible might encourage more competition and improve the industry's efficiency. This and other decisions taken in the context of the WTO might have a significant impact on the space sector, if governments decide they want to foster commercial space activities.

Export controls tend to restrict the ability to exploit market opportunities

In addition to licensing requirements, space firms may also have to contend with export controls when they attempt to market their products and services abroad. In this regard, the US ITAR regime (Box 5.7), which applies to the transfer abroad of sensitive technologies, may have adversely affected the ability of US space firms to tap foreign markets. In practice, ITAR regulations concern not only commodities, but also the transfer of technical information (*e.g.* written documents, oral presentations). This means that private firms and public space-related organisations need to apply not only for export licences to ship certain dual-use items to customers or partners organisations, but also – when discussing projects with foreign third parties in meetings and conferences – for specific technical assistance agreements and even, in certain cases, Department of Defense monitoring. Especially in the case of telecommunications satellites, several US firms (*e.g.* satellite manufacturers, component suppliers and other exporters of technology products) have complained about the rather lengthy administrative procedures since 1999; they feel at a disadvantage compared with their European or Asian competitors, as they have experienced serious delays in approval of exports and also prohibitions on exports (de Selding, 2004a).

> Box 5.7. **The US regime for technology transfer: International Traffic in Arms Regulations (ITAR)**
>
> The United States has a relatively strong regulatory regime to prevent the illegal transfer and theft of sensitive technologies, such as space systems that can be used in the development of military assets by governments, entities and persons that may be hostile to American interests. Authorisation from the US government must be secured, sometimes following lengthy procedures, to export or re-export satellites and most satellite components to a foreign country, or to launch a satellite on a foreign launch vehicle:
>
> - The US Arms Export Control Act (AECA) of 1976 is the primary law establishing procedures for sales and transfers of military equipment and related services.
>
> - The International Traffic in Arms Regulations implement the EACA law and guide defence trade activities. The ITAR text defines in particular which items are to be considered as munitions (Section 121 – the United States Munitions List) with different categories of systems considered.
>
> - Since most space technologies are dual use, some items originally designed, developed or manufactured to military specifications are subject to the ITAR, even if they are made available for civil or commercial use. This includes different types of space systems and components such as launch vehicles (e.g. space launchers belong to category IV of the US Munitions Lists: Launch vehicles, guided missiles, ballistic missiles, rockets, torpedoes, bombs and mines).
>
> - The Department of State has responsibility for developing and updating the ITAR regime, including managing licences and the diverse authorisation procedures linked to defence trade (e.g. amendments regularly made to remove certain countries from the list of proscribed destinations for the exports and imports of US defence articles and services).
>
> Source: US DoC (2004); de Selding (2004a).

The application of ITAR also extends to non-US firms using US space components if such components are deemed sensitive under the export control regime. For instance, this may prevent European satellite manufacturers from exporting to China, if the spacecraft to be exported include US components subject to ITAR. This extension of the ITAR regime to non-US firms has had unintended consequences in other countries. It has induced a number of space systems manufacturers in Europe and Asia to develop new technologies in order to reduce their dependence on US-made components and create ITAR-free products, with some of them lobbying for a possible "Buy European Act". A recent example is the development of Pleiades, the upcoming French dual-use high-resolution optical imaging system. The satellites to be built by EADS Astrium and Alcatel Space and launched in 2008 will not have any US components that are on the State Department's ITAR list. In economic terms, these policies have in fact accelerated import substitution in many countries and defeated the intent of ITAR.

Hence, while it is obvious that legitimate security concerns need to take precedence over commerce, it is also clear that technology transfer constraints tend to have an economic cost by dampening the exports and imports of space services and products. Moreover, such regimes encourage others to "innovate around the constraints", a potentially wasteful process from a global perspective.

It is generally government policy and the myriad levels of approvals necessary to conduct commercial activities rather than space laws that discourage some space business activities in space-faring countries. Because of governments' treatment of space as a security and technology demonstration programme, rather than as a business opportunity, the levels of involvement and review by public agencies are even more stringent than for most other economic sectors.

In recent years, the attitude of governments has changed somewhat and the economic dimension of space has been increasingly recognised. For instance, in the United States, almost every Presidential Space Policy Directives of the past 15 years has called for more private-sector involvement in space. However, these policy directives make clear that national security is still the prime concern and that private activity will have to yield to security concerns. Defining a security concern is purposely left open, making these policies difficult to interpret. This is also true of other countries where the monopolistic position of governmental agencies in space affairs often tends to discourage the development of new actors.

The space debris issue is inadequately addressed

Space debris, composed mostly of man-made space objects that have fallen apart or are out of order (*e.g.* satellites, missile components), may seriously hamper future space activities in orbit, space travel and scientific research. Scientists estimate that there are already more than 100 000 objects between 10 cm and 1 cm in size that are too small to be verifiably detected and followed with current technology, and perhaps trillions of smaller pieces (Hitchens, 2004). However, no major plan has been put into action to clear the debris – if that is even possible – although some steps have been taken to avoid the creation of further debris.[12]

A 1993 ITU recommendation on the environmental protection of the geostationary satellite orbit (GSO) urges minimising debris released into the GSO region during the placement of a satellite in orbit, and also

> Box 5.8. **The Inter-Agency Space Debris Co-ordination Committee**
>
> The IADC, set up in 1993, is an international governmental forum for the worldwide co-ordination of activities related to the issues of man-made and natural debris in space. Eleven space agencies are members.* The primary purpose of the IADC is to exchange information on space debris research activities among member space agencies, to facilitate opportunities for co-operation on space debris research, to review the progress of ongoing co-operative activities and to identify debris mitigation options.
>
> In 2002, the IADC submitted to the United Nations' Committee on the Peaceful Uses of Outer Space (COPUOS) a set of guidelines regarding the mitigation of space debris. The IADC guidelines ask countries to limit debris released during normal space operations, minimise the potential for on-orbit break-ups, undertake post-mission disposal and prevent collisions. In addition, IADC recommends that a space debris mitigation plan be put together for each space project, and asks nations to report voluntarily (beginning in 2005) on mitigation efforts. Experts hope that these guidelines can be agreed at the next meeting of the COPUOS Science and Technical Subcommittee in 2005, as a first tentative step to an international space debris mitigation regime.
>
> * ASI (Agenzia Spaziale Italiana); BNSC (British National Space Centre); CNES (Centre National d'Études Spatiales); CNSA (China National Space Administration); DLR (German Aerospace Centre); ESA (European Space Agency); ISRO (Indian Space Research Organisation); JAXA (Japan Aerospace Exploration Agency); NASA (National Aeronautics and Space Administration); NSAU (National Space Agency of Ukraine); ROSAVIAKOSMOS (Russian Aviation and Space Agency).
>
> Source: IADC (2004); Hitchens (2004).

the transfer of the GSO satellite at the end of its life to a graveyard orbit (Recommendation ITU-RS.1003-1). The recommendation has been endorsed by the Inter-Agency Space Debris Co-ordination Committee (IADC), which has developed other guidelines to mitigate the development of human-made space debris (Box 5.8).

As a result, some "good practices" concerning space debris are slowly being put in place by major space-faring countries, nationally and through the IADC. They include, in particular, regulatory standards aimed at limiting the creation of debris from government-sponsored space operations.[13] A few commercial space operators are tentatively following these practices, but they are still few in number, as extra resources are needed when trying to avoid the creation of debris (*e.g.* extra fuel necessary to dispose of satellites in safer orbits when their mission is over).

Although it would be useful to increase international co-operation on this issue and create a coherent regime, based on the IADC guidelines for instance, there will always be a problem for identifying debris and probable causes of accidents involving debris.[14] Indeed, accidents with space debris are not comparable to terrestrial accidents for which an investigation is reasonably likely to determine fault conclusively.

Although little progress has been achieved so far at international level, positive steps have recently been taken, in particular in the United States. Over the objections of several of the world's largest commercial satellite fleet operators, the Federal Communications Commission (FCC) ruled recently that, once they are no longer in active use, all US-licensed satellites launched after 18 March 2002, will have to be placed into so-called graveyard orbits between 200 km and 300 km above the geostationary arc where most commercial satellites operate (de Selding, 2004b). The FCC based its new rules on recommendations made by the ITU and the IADC.

Several standardisation questions remain open

Despite the global standardisation process under way, some applications-specific standardisation issues need to be addressed. In the analysis of specific space applications in the third phase of the project, the key role of standards was clearly demonstrated, notably for ensuring system scalability, fostering interoperability and competition among commercial providers, and encouraging user uptake. It was found in particular that each field of application has – besides some obvious commonalities – some very specific requirements in terms of interoperability and standards (e.g. remote sensing vs. telecommunications particularities) that need to be dealt with. Moreover, since space-based applications are being developed across a large range of non-space disciplines that already have their own standards (e.g. telehealth: management standards, health standards, data transfer standards), there is a need to make the entire system as coherent as possible.

Government involvement in international standards activities in relation to the use of space is critical. Industry cannot take on this role alone through the normal commercial process, as a significant share of international space infrastructure is publicly owned and space agencies play a key role in the development of space technologies. Moreover, it is in the best interests of public actors to play an active role in standards setting, as the public sector is the main user of space products and services. Any improvements in the efficiency and interoperability of public systems will have a substantial impact on governments and their taxpayers.

5. FRAMEWORK CONDITIONS: LEGAL, REGULATORY AND PUBLIC AWARENESS ASPECTS

Public awareness aspects

Public awareness is crucial for the political sustainability of the space sector and for maximising the socio-economic benefits from space activities. Unfortunately, such awareness is limited. Although some space ventures have attracted media attention over the years, the general public generally lacks a genuine understanding of the concrete contribution of space applications and, consequently, does not fully appreciate the value society at large receives from space activities. This adversely affects the ability of decision makers to take appropriate action for the development of space systems in a timely manner and also affects negatively the preparation of future generations of space scientists and engineers.

What has been achieved

Historically, significant space exploration achievements have attracted public interest worldwide. In the 1960s, the Apollo landing on the Moon captured the imagination of many, and inspired generations of scientists and engineers. The Mars Pathfinder mission, with Mars pictures directly downloadable on personal computers, motivated a large number of students to study exobiology (an interdisciplinary scientific field which studies the origin, evolution and distribution of life in the Universe) and become "planet hunters". More recently, NASA's Mars rovers Spirit and Opportunity and the European Mars Express mission have triggered public enthusiasm.

Major space-related policy decisions also tend to attract public attention. The decision to launch "a race to the Moon" between the United States and the Union of Soviet Socialist Republics in the 1960s had at the time an impact on international public opinion. Although no follow-up challenge was able to catch the public imagination to the same extent once the race was over, there was growing awareness among the general public of the strategic as well as scientific dimensions of space assets.

More recently in Europe, the Green and White Papers formulating a new Europe-wide space policy have involved many actors in the space community, but also non-space players, thereby opening up a traditionally closed domain to outsiders. This has created increasing interest in various sectors that may benefit from space systems. Another parallel policy effort, the inclusion of space in the European Constitutional Treaty, has given more prominence to space, in the eyes both of decision makers and of the general public.[15]

In the United States, President Bush's space initiative, an ambitious plan to return Americans to the Moon by 2020 and use the mission as a stepping stone for further human trips to Mars and beyond, has received a great deal of media attention.

Aside from such highly visible political decisions, space entrepreneurship is also becoming a new vector to spur interest in the space sector. Tales of daring enterprises tend to capture the imagination of the general public. This is not new: the first car races and the first airplane flights excited large crowds. Recently similar excitement was generated by the X-Prize competition, which was won in October 2004 by SpaceShipOne. Though the competition was generally criticised at first, in part by people active in the space sector itself, the prize has given the general public a positive vision of dynamic space companies and has attracted the interest of entrepreneurs outside the space sector.[16]

Issues to be addressed

While space ventures have indeed received a fair amount of attention in the past, the general public has a somewhat distorted vision of what space can offer. This is because, in the information age, the general public and decision makers alike are overwhelmed by an excess of data, news, sports, sciences, etc., while important but not spectacular space-related news does not get the attention it deserves. A major challenge in coming years will therefore be to foster public understanding of the value of space activities and to address the lack of interest in space careers among students.

Poor public understanding of the value of space

Space programmes usually require a good level of public support if they are to generate large benefits. A virtuous circle can be attained when the public is interested in projects and policy makers take decisions that respond to their constituencies' wishes. This is not often the case for space. Indeed, public spending on space has often been criticised on four main grounds.

One is the large budgets – at least in the public eye – allocated to space, without enough clearly visible returns to society. Many people would rather see public authorities invest more in programmes to fight poverty, instead of focusing on technology developments they do not relate to. The main problem is that satellite applications are mostly "background" applications akin to utilities. People are generally unaware that they use space services (e.g. telecommunication satellite backbone links), or they take such services for granted (e.g. GPS), even though a disruption in such services can have far-reaching implications.[17]

Second is the apparently routine nature of many space activities nowadays, although it is still very risky to launch payloads to orbit. Regular successes, however, rarely make the news, and the general public will more likely be made aware of launch failures, as in the case of the extensive media coverage of the tragic Columbia accident in 2003.

Third is the running of public agencies themselves. Budget overruns tend to discredit space agencies' achievements in the eyes of the public. For instance, although NASA's budget is only on average 1% of the overall US federal budget, heated yearly discussions in Congress regarding the space agency's programme costs overruns, especially in the case of the International Space Station, damage NASA's reputation.[18]

Although the management problems faced by NASA are real, it is important to keep in mind that running space agencies, notably large ones, is very challenging. Politicians and the general public tend to forget that leading-edge activities are hard to manage financially and are by nature risky: development costs are difficult to estimate in advance, and costs overruns are to be expected. Moreover, risks of failure exist. Indeed, a space programme without failures would be suspect: it would mean that it is not ambitious enough. Science and technology do not progress without the lessons learned from failures.

Fourth, concerns are likely to rise in the coming years about the intrusive nature of space technology, especially regarding commercial Earth observation activities and location-based services using navigation satellites. High-resolution imagery downloadable from the Internet and location sensors integrated in much electronic equipment may spur a negative reaction from segments of the population and businesses in OECD and non-OECD countries alike, as fears about loss of privacy and confidentiality breaches increase.[19]

It follows from these considerations that there is a significant information gap between perception and reality. What appears to be lacking is a fair and balanced assessment of the costs and benefits of space applications that mainstream policy makers and the general public can relate to. Such an assessment is particularly important in the case of applications that have a strong public good orientation and are thus unlikely to be developed without public support.

Lack of interest in space careers

Space systems can provide many socio-economic benefits, but a qualified workforce able to maintain existing systems and develop innovative ones is essential to the future of the sector in all countries. According to Daniel Goldin (2004), former NASA administrator, "the most important societal impact of space and aerospace technologies remains that they further human inspiration". Without this inspiration, the future space workforce will emerge only with great difficulty.

In 2002, the European Science Foundation (ESF) led a study showing that different countries have different problems in terms of their space-related workforce (ESF, 2003). However, one main element emerged clearly, namely the sharp decrease of the population under 25 years old in the space science workforce, including engineers and scientists, all over Europe.

This trend was linked in part to the diminishing attractiveness of scientific studies among younger generations. It is also linked to a decline in interest in space among scientifically minded students. Indeed, many future scientists and engineers look at other sectors for technical challenges or better salaries (*e.g.* software development, biotechnologies).

The ESF made several recommendations to European ministries of research, advocating workforce mobility among member states facing different situations, and stressing the need to promote space-related education programmes at the undergraduate level.[20] The same problem exists in the United States. For instance, NASA's workforce under 30 years of age is one-third the size of its workforce over 60, and 25% of its scientists and engineers will be eligible to retire in the next three to five years.

A major difficulty for attracting large numbers of qualified workers in space-related careers is the relative lack of good job opportunities in the sector, when compared to other high-technology sectors. Rationalisation efforts in the industry worldwide in the 1990s and lower governmental budgets in some cases have concentrated expertise. Moreover, cyclical staff layoffs have discouraged many would-be candidates. In this context, it may be misleading and wasteful to encourage young people to pursue a career in the space sector, if they are likely to have a hard time finding a job (whether in public research centres or industry), once they have completed their studies. Yet, for the future of the sector, it is crucial to maintain a critical pool of expertise.

Some large space and defence firms have put in place new recruiting policies, requiring employees to be more mobile and able to work on different dual systems when needed (*e.g.* moving from civil space projects to military ones). This approach might not work for recruiting specialists in specific scientific disciplines or experts on particular technological subsystems.

Another approach is to try to increase the attractiveness of space careers. For instance, a bill was recently passed in the US Congress to allow more benefits to future NASA employees in order to attract more qualified candidates, although there are few jobs available in space agencies at this time.[21] Despite these and other measures, many in the sector fear that "band-aid" programmes to encourage new entrants may have an effect at the margin but will not go to the core of the problem.

It follows that sustained efforts are needed to raise public awareness of both the costs and the societal benefits of space programmes and to encourage new generations of engineers and scientist to chose space-oriented careers. Without such efforts, the political will to develop new systems for meeting societal needs could be undermined, while failure to renew the space workforce could have devastating consequences in terms of losses of skills and expertise which have taken decades to build.

Notes

1. According to international space law, governments are responsible for the space activities carried out by national public agencies and by non-governmental entities operating on their territory. Activities in outer space, including those involving the Moon and other celestial bodies, require authorisation and continuing supervision by the responsible state (*i.e.* licence to operate).
2. Examples include: United Kingdom (Outer Space Act, 1986); United States (NASA Act, 1958; Commercial Space Launch Act, 1984; Communications Satellite Act, 1962; Land Remote Sensing Policy Act, 1993); South Africa (Space Affairs Act, 1993); Russian Federation (Law on Space Activities, 1993; Statute No. 104 on licensing space operations).
3. This aspect was further discussed in seminars held in 2003 and 2004 at the French National Assembly, which concluded that there was a clear need to elaborate a domestic legal framework for space activities, with a proposal in that sense for a draft legal text in 2004-05.
4. Project 2001Plus is the follow-up to Project 2001 which covered some basic issues concerning space commercialisation. The aim of Project 2001 Plus is to analyse the consequences of globalisation and European integration for future commercial space activities. One workshop, "Towards a harmonised approach for national space legislation in Europe", held in January 2004, aimed at developing common structures for harmonising national space legislation in Europe.
5. See in particular the Project 2001 and the Project 2001 Plus.
6. Among the current UNIDROIT projects, the UNIDROIT *Convention on International Interests in Mobile Equipment* was opened, after several years of negotiations, for signature in Cape Town on 16 November 2001. The Convention is split into a base Convention setting rules universally applicable for the creation, registration and enforcement of security to different categories of equipment and several equipment specific protocols containing additional rules adapted to the special financing patterns of those particular categories of equipment. The mobile equipment under consideration include: *a)* airframes, aircraft engines and helicopters; *b)* railway rolling stock; and, *c)* for the first time space assets. A first protocol on aircraft was opened for signature at the same time as the General Convention in 2001. So far, 28 states have signed the Convention and its related dedicated Protocol on Matters Specific to Aircraft Equipment (October 2004).
7. As a positive step in that direction, the second session of the UNIDROIT Committee of Governmental Experts to prepare the draft protocol on space assets, held in October 2004, established an open-ended subcommittee to develop proposals relating to the international registration system to underpin the future protocol (UNIDROIT, 2004b).
8. The Space Working Group plays an important role in the work of the Committee of Governmental Experts, ensuring that the viewpoint of the different industry sectors is taken into consideration.
9. One of the fundamental characteristics of the TRIPS Agreement is that it makes protection of intellectual property rights an integral part of the multilateral trading system, as embodied in the WTO. The TRIPS Agreement is often described as one of the three "pillars" of the WTO, the other two being trade in goods (the traditional domain of the GATT) and trade in services. It applies to all WTO members and the provisions of the agreement are subject to the integrated WTO dispute settlement mechanism which is contained in the Dispute Settlement Understanding (the "Understanding on Rules and Procedures Governing the Settlement of Disputes").

10. The Minneapolis Plenipotentiary Conference, held in 1998, introduced for the first time an administration and processing fee for all new systems (Resolutions 88 and 91) in order to dissuade paper satellites, although the fee is extremely low in relation to total system costs. The "due diligence" principle obliges all operators and administrations filing system co-ordination requests to provide full details of system contractors, including manufacturers and launch companies, along with a planned schedule of system deployment. The 2002 ITU Plenipotentiary Conference, held in Marrakech, Morocco, enacted a cost-recovery resolution to impose penalties (i.e. loss of "place in line" in the ITU priorities queue) for non-payment of cost recovery fees. This resolution should help reduce the satellite network filings backlog by weeding out at least some "paper satellites".

11. This resolution invites the Radiocommunication Sector of the ITU to examine the effectiveness, appropriateness and impact of the Radio Regulations, with respect to the evolution of existing, emerging and future applications, systems and technologies, and to identify options for the improvement of the regulatory regime.

12. Man-made debris orbits at a speed of roughly 17 500 miles/hour (28 000 km/h). Even an object as small as a grape has enough kinetic energy to permanently damage a medium-sized spacecraft.

13. These good practices include: limiting the amount of debris produced from normal operations, such as throwaway orbital stages or components; burning off fuel at the end of a satellite's mission life; removing non-operational spacecraft and rocket stages from orbit, either by de-orbiting objects in low Earth orbit (over a certain time) or boosting them up and out of the way into a so-called graveyard orbit for objects in geostationary orbits (IADC, 2004).

14. Article VIII of the Outer Space Treaty provides that the state of registry retains jurisdiction and control of an object launched into outer space while that object is in space. But whether that state is responsible for clearing up "its" space debris and how to realistically enforce any obligations for that state to do so is still an open issue.

15. The European Union (EU) Constitutional Treaty still needs to be ratified by the 25 member states before entering into force. Its Article 13 grants the EU shared competence in space, which should give the EU a stronger role in defining and implementing the common European Space Policy with member states.

16. The X-Prize was a private initiative that led private entrepreneurs to make substantial efforts to develop a suborbital plane with no or marginal government support. The prize was won in September 2004 by SpaceShipOne, the first private suborbital plane to reach space. This success and the efforts of other space entrepreneurs could pave the way for the development of a job- and revenue-generating space adventure and tourism industry in the coming years. These space entrepreneurs might also contribute, even if only in modest ways, to the development of new flight technologies and cost-efficient operations mechanisms.

17. For instance, a poll conducted after the announcement of President Bush's space initiative indicated that, given the choice of spending money on programmes like education and health care or on space research, 55% of respondents said they wanted domestic programmes (Associated Press, 2004).

18. The US Government Accountability Office (GAO) began publishing in 1990 a list of "high-risk programmes", i.e. federal programmes vulnerable to fraud, waste, abuse and mismanagement owing to the way they are run. NASA contract management has made the high-risk list every time since 1990, although the GAO notes that NASA has made some improvements this year. It has notably made progress in the implementation of a single financial management system for all ten of its field centres (Berger, 2005).

19. These issues are addressed in a broader context in OECD (2004). This study examines the main features of the emerging security industry and security economy in OECD countries, and its far-reaching economic and social implications.

20. The ESF study mentions in particular the following reasons for the lack of interest in scientific studies and careers (based on an opinion poll aimed at European young people still studying in 2001 in EU member states): lack of appeal of scientific studies (67.3% of respondents); difficulty of the subjects (58.7%); young people are not so interested in scientific subjects (53.4%); salaries are not attractive enough (40%); science has too negative an image (34%).

21. The NASA Work Force Flexibility Act of 2003 (HR1085 and S.610), passed January 2004, gives NASA additional powers to hire and retain qualified engineers and scientists.

Bibliography

Associated Press, "Public Tepid on Bush Space Plan", *Associated Press Release*, 13 January.

Becker, F. (2002), "Space Education: The International Space University Model", *Online Journal of Communications*, Issue 1, Spring.

Berger, B. (2005), "NASA Contract Management on GAO's High-Risk List", *Space News*, 31 January.

Cheng, B. (1998), *Studies in International Space Law*, Oxford University Press, Oxford.

ESF – European Science Foundation (2003), *Demography of European Space Science: Results from an ESSC-ESF Study*, Strasbourg, France, April.

Euroconsult (2004), *World Prospects for Government Space Markets: 2004 Edition*, Paris.

Foreign Affairs Canada (2004), "Canada Tables Legislation Regulating Remote Sensing Space Systems", *Foreign Affairs Canada*, No. 136, 23 November.

Goldin, Daniel S. (2004), "Bold Missions and the Big Picture: The Societal Impact of America's Space Program", *Technology in Society*, Vol. 26, No. 2/3.

Hitchens, T. (2004), "Space Debris: Next Steps", Center for Defense Information (CDI), 1 April.

IADC – Inter-Agency Space Debris Coordination Committee (2004), *www.iadc-online.org/*, accessed September 2004.

ILA – International Law Association (2002), "Space Law Committee Final Report on the Review of Space Law Treaties in View of Commercial Space Activities – Concrete Proposals", Professor Maureen Williams, General Rapporteur, ILA Conference, New Delhi.

ITU – International Telecommunications Union (2003), "Paper Tigers: The Scramble for Space Spectrum", ITU *Press Release*, *www.itu.int/newsarchive/pp02/media_information/feature_satellite.html*, accessed 10 July 2004.

Malanczuk, P. (1999), "The Relevance of International Economic Law and the World Trade Organization (WTO) for Commercial Outer Space Activities", *Proceedings of the Third ECSL Colloquium*, ESA-SP-442, ESA, Noordwijk, May.

Ministère délégué à la Recherche et aux Nouvelles Technologies (2002), *L'évolution du droit de l'espace en France*, 12 November, Paris.

OECD (2004), *The Security Economy*, OECD, Paris.

Reif, S.U. (2002), Shaping a Legal Framework for the Commercial Use of Outer Space: Recommendations and Conclusions from Project 2001, *Space Policy*, pp. 157-162.

de Selding, P. (2004a), "US Export Restrictions Help Alcatel Win Chinasat", *Space News*, 9 June.

de Selding, P. (2004b), "FCC Enters Orbital Debris Debate", *Space News*, 28 June.

UN OOSA – United Nations Office of Outer Space Affairs (2004), *www.oosa.unvienna.org/SpaceLaw/treaties.html*, accessed 11 August 2004.

UNIDROIT – International Institute for the Unification of Private Law (2004a), "UNIDROIT's Project for the Enhancement of Access to Commercial Space Financing", Report prepared by Martin Stanford, Principal Research Officer, for the OECD Project on the Commercialisation of Space and the Development of Space Infrastructure, Third Meeting of the Project's Steering Group, 7 May.

UNIDROIT (2004b), "Report of the Second Session of UNIDROIT Committee of Governmental Experts for the Preparation of a Draft Protocol to the Convention on International Interests in Mobile Equipment on Matters Specific to Space Assets", Rome, 26-28 October, C.G.E./Space Pr./2/Report.

US DoC – US Department of Commerce (2004), Bureau of Industry and Security, *www.bxa.doc.gov*, accessed 4 October 2004.

Van Traa Engelman, H.L. (1993), *Commercial Utilization of Outer Space, Law and Practice*, Martinus Nijhoff Publishers, Dordrecht.

Von der Dunk, F.G. (1998), "Private Enterprise and Public Interest in the European "Spacescape' – Towards Harmonized National Space Legislation for Private Space Activities in Europe,", International Institute of Air and Space Law, Faculty of Law, Leyden University, Leyden.

Von der Dunk, F.G. (1999), "International Organisations and Space Law – Their Role and Contribution, 3rd ECSL Colloquium, 6-7 May 1999, Perugia", *Air and Space Law*, Vol. 4, Issue 3, pp. 164-168(5), July.

Von der Dunk, F.G. (2004), "Heeding the Public-Private Paradigm: Overview of National Space Legislation around the World, in 2004", *Space Law Conference Papers Assembled*, pp. 20-34, University of Leyden, Leyden.

WIPO – World Intellectual Property Organisation (2004), "Intellectual Property and Space Activities", Issue paper prepared by the International Bureau of WIPO for the OECD Project, April.

Chapter 6

Main Findings and Recommendations

This chapter summarises the most important lessons learned throughout the project and draws from these findings a number of recommendations for the attention of decision makers. In a nutshell, the main conclusion is that the space sector has indeed a promising longer-term future, but that this potential will not be fulfilled unless governments take decisive action to improve the framework conditions governing space activities. This provides the basis for outlining in the balance of the chapter three sets of complementary measures that should help ensure that socially useful space solutions are developed and contribute fully to addressing major societal challenges in coming decades. These include: i) measures designed to implement an efficient, robust and sustainable space infrastructure; ii) measures to foster greater public use of space solutions, when it is cost-effective to do so; and iii) measures to encourage private-sector actors to participate more fully in the development and operation of space systems, by creating a more business-friendly environment for space at both national and international levels.

Lessons learned during the project

A number of important lessons have been learned throughout this project. The first is encouraging: the longer-term future of the space sector is promising over a broad range of global scenarios, when all main segments of the space sector (i.e. military space, civil space and commercial space) are taken into account.

However, it is somewhat tempered by the second lesson which suggests that severe short- and medium-term fluctuations are likely to affect space actors, given the capital-intensive nature of space activities, the long lead times required for the development of space assets, the high risks of space ventures and the heavy involvement of the state in space activities.

It also appeared that a clear distinction should be made between the upstream segment of the sector (space asset manufacturing and launching services) and the downstream segment (space applications). Typically, the downstream segment offers better prospects over the longer term than the upstream segment which suffers from a situation of chronic oversupply resulting largely from the desire of governments of space-faring nations to establish and maintain – for strategic and national sovereignty reasons – independent (if not guaranteed) access to space.

However, while the downstream sector offers the best prospects overall, not all applications are equally promising over the 30-year period considered. On the one hand, information-intensive applications such as satellite-based telecommunications, Earth observation and navigation have a bright future. On the other, the prospects of transport and manufacturing applications are more uncertain, given the cost of access to space, which is unlikely to decline drastically over the period, and the complex technical problems of working in space.

The third lesson learned is that space can help cope with a number of major societal challenges that will confront humankind in coming decades. These range from serious threats to the physical environment (climate change, growing pollution, depletion of natural resources and the impact of intensive agricultural practices) to major social challenges (evolution towards the knowledge society, increased mobility and rising concerns about security).

For each of these challenges, it was found that space can make a useful contribution for both OECD and non-OECD countries.

Space and societal challenges

Regarding *environmental challenges*, space infrastructure – composed of Earth observation (EO) and navigation systems – provides data that can be used for weather forecasts as well as for assessing greenhouse gas (GHG) emissions, monitoring air pollution, detecting potential anthropogenic change, validating climate models and predicting future change. It can also be used for monitoring changes in the natural environment, such as the evolution of fault lines, landslides, subsidence and volcanoes. Moreover, space-based systems might also be used for monitoring the application of Kyoto commitments.

Regarding *challenges for managing natural resources and agriculture*, space-based data have a broad range of applications. First, regarding energy, such data provide information on both current and future states of the energy system and the environmental context. They can also be used for controlling power and pipeline distribution systems, hydropower dam operation and wind power generation. Second, Earth observation data facilitate the management of water resources through better understanding of the water cycle, notably by providing information on atmospheric temperature and water vapour, sea surface temperatures, ocean winds, 3-D information on rainfall structure and characteristics, soil moisture and ocean salinity. Third, space technology is useful for managing forest resources more effectively and combating deforestation. Remote sensing data also provide useful information about the aerial extent, conditions and boundaries of mangrove forests and have proved extremely useful for wetlands mapping and for determining high and low water lines. Finally, space systems have also important application in agriculture when combined with other technologies. Global navigation space systems (GNSS) and spaced-based augmentation systems (SBAS), geographic information systems, miniaturised computer components, automatic control and in-field and remote sensing can be used to appraise the state of crops, identify areas requiring attention and target treatment automatically.

Regarding *security challenges*, the capacity of space-based systems to see, locate and communicate over broad areas finds a growing range of applications. For instance, space systems can provide useful input to disaster management information systems throughout the disaster management cycle. The Global Navigation Satellite System (GNSS) allows first responders to quickly pinpoint the scene of an accident, thereby reducing response time for emergency services, while space-based telemedicine applications can enhance the ability of emergency personnel to treat victims quickly and effectively. GNSS can also be used for tracking and controlling the transport of illegal and hazardous goods. Moreover, space-based systems can be used for monitoring compliance with international treaties and for the surveillance of international borders.

Regarding *mobility challenges*, space-based systems can be used for a broad range of traffic management applications, including route guidance (selection of optimum route in real time), the management of traffic flows (monitoring of traffic flows in real time, anticipation of traffic jams and implementation of remedial action in real time), fleet management, advanced driving assistance systems and road-charging schemes. Air traffic control represents another major area of application of space-based augmented systems.

Regarding *challenges related to the move to the knowledge society*, space plays a dual role. First, the R&D efforts of space agencies and other space actors create new knowledge that can be applied both in the space sector and in other sectors of the economy. Moreover, Earth observation and deep space missions generate an unprecedented wealth of data and information on the state of our planet and of the Universe. Second, space facilitates the distribution of knowledge: satellite communication is an essential element of the communications infrastructure. Satellites have specific advantages in terms of international coverage, broadcasting, flexibility and rapid deployment of service. They have been very successful in some market segments, such as direct broadcasting satellites (DBS) and help foster competition and innovation in those markets. They also provide the technical means for the delivery of some public services (*e.g.* in rural and remote areas, for emergency services). The role of space for the distribution of knowledge – and more generally for communication – is particularly important in developing countries, where, typically, the ground-based infrastructure is limited or inexistent.

There are good reasons to believe that, in the future, the five challenges outlined above will increase in importance, making the potential contribution space can make to addressing them even more valuable, especially in light of the expected advances in space technology. For example, on the environmental front, human activity, notably the burning of fossil fuels, will have significant consequences for the world's climate in the coming decades. This will call for ever more stringent emission abatement measures, if life on Earth as we know it is to be preserved. Similarly, without appropriate action, excessive use of natural resources (including water and tropical forest resources), as well as intensive forms of farming can result in drastic reduction in biodiversity, threats to food security and, more generally, irreversible damage to life on Earth. Security concerns, too, will require greater attention. Partly because of climate change and loss of biodiversity, partly because of increasing urbanisation, industrialisation and the growing interdependence of systems, partly because of the threats linked to global terrorism and organised crime, disasters – whether natural or man-made – are likely to become more frequent and more costly in terms of lost lives and damage to property.

Moreover, while growing mobility will bring about substantial economic and social benefits, it will also impose an ever-increasing burden on society at large over the coming decades. Economic development will cause unsustainable increases in the volume of traffic on the roads, in the air as well as on the seas, with significant externalities in terms of pollution, emissions of greenhouse gases, congestion and loss of lives and property in accidents. And finally, as the information revolution continues to unfold, knowledge-based activities will increasingly dominate the economy, calling for the development of an information infrastructure capable of bringing to all citizens a growing range of electronic services.

The role of governments

While space's potential contribution to addressing society's future challenges can only increase, it cannot be taken for granted that this potential will actually be realised. Much will depend on public policy and the framework conditions that govern space activities, given the dominant role played by governments in the sector. Questions arise regarding the development of space infrastructure and the role to be played in this regard by public and private actors. Moreover, the success or failure of space applications is affected not only by space policy *per se* but also by general laws and regulations (*e.g.* economic, social and environmental policies) and their application.

Case studies conducted during the third phase of the project underscore the critical role of these general laws and regulations. They also suggest that there are significant commonalities across applications in terms of their impact. First, the importance of a stable and predictable legal and regulatory environment was evident in all cases studies. Other strong messages are the need to deal effectively with uncertainties that relate to liability, notably for emerging applications, and the importance of creating and preserving a balanced competitive environment when the services provided by a given application have to compete with services offered by other actors.

Equitable access to services was another major theme, one which extends beyond the digital divide between rural and urban dwellers to encompass questions of equal treatment of individuals and national entities regarding access to information and knowledge derived from space activities.

In most of the case studies, issues relating to the generation, distribution and use of information also played a prominent role, raising questions concerning intellectual property, the pricing of data and the problem of data confidentiality and privacy.

The case studies also demonstrated that greater compatibility of technological systems, standards, licensing practices and so on are central to the future development of space applications. Moreover, the key role of infrastructure and the extent to which public authorities should be involved in its provision and operation are recurring issues. Finally, in a number of instances, there was a clear-cut case to be made for encouraging government support of R&D.

While the role of government is essential for the development of space applications, the conditions under which public authority is exercised today is far from ideal. This applies in particular to the framework conditions (i.e. the existing institutional, legal and regulatory regime) that determine how society at large is organised for meeting the challenges of the future.

First, there are a number of uncertainties on the institutional front regarding how the different public and private space actors are positioned to discharge their responsibilities, the relationships that obtain among them and the incentives in place that shape their behaviour. These uncertainties concern in particular:

- The role and place of space agencies in government (*e.g.* Who should they report to? Should they be involved in the running of applications?).
- The relationship between space agencies and user departments (*e.g.* How can an effective dialogue between providers and users of space services be established?).
- The role of public and private players (*e.g.* Who should do what? How should space agencies help private actors? How should they co-operate?).
- The dual use of the technology (*e.g.* What kind of co-operation/control should the military have with/over civil agencies? With/over private actors?).

A second set of issues on the legal and regulatory front relates to the rules of the game that space actors have to abide by. Major problems facing space actors result from the lack of national laws in a number of space-faring countries; from the fact that existing space-related laws are not business-friendly; from the fact that in many applications, success or failure depends on the application of laws beyond space laws (*e.g.* liability, copyright); from the fact that the international law regime is not very well-suited to business; from the limited application of World Trade Organisation disciplines to trade in space goods and services; and from regulatory issues raised by the operation of the International Telecommunications Union.

Finally, it was noted that although some space ventures have attracted a lot of public interest, there is a lack of awareness in the general population of the concrete contribution that space can make to society at large, with adverse consequences for the political decision process. As a result, decisions regarding space are not always taken with a full understanding of the issues at hand. Moreover, the lack of attractiveness of space-related careers is a source of concern for the future of the sector. There is a danger that the knowledge and expertise accumulated over decades may be lost in the coming years if too few students are attracted to careers in the space sector.

Given the central role they play, only governments – individually and collectively – are in a position to remedy the shortcomings noted above.

Purpose, scope and overall architecture of the recommendations

The recommendations drafted on the basis of these findings are intended to provide a long-term, future-oriented framework, i.e. an overall, consistent set of broad policy orientations that can offer a useful framework for policy formulation. The recommendations are made from a broadly societal non-space perspective and are therefore addressed to governments in general, rather than to the space community as such. More specifically, they are intended for ministries that have main responsibility for overall economic and social policies – including policies that may have a bearing on the performance of private space actors – as well as for user departments that can take advantage of space-based solutions for delivering their services to the general public.

The recommendations focus on the "big picture" and take a long-term policy view. They aim to address what governments can do to strengthen the contribution that space can make to the solution of the major socio-economic challenges to be faced over the coming decades. Moreover, they extend beyond the traditional ambit of space policy *per se* to other policy areas that may have a bearing on the successful deployment and use of space applications for meeting societal challenges, although the range of issues covered is by no means exhaustive.

The recommendations are constructed with a "bridge" in mind: How do we get from "here and now" to a much-improved situation 20 years or so down the road? The "surface" of the bridge consists of three blocks of roughly equal importance that stand for a cluster of policies for achieving a specific but broad-based objective, namely:

- **Block I:** Measures to implement a space infrastructure that is sustainable, that fully takes into account user needs, and that is fully integrated with complementary ground-based infrastructure.
- **Block II:** Measures to take advantage of the productivity gains that space solutions may offer for the delivery of public services and the development of new ones.
- **Block III:** Measures that encourage the private sector to contribute fully to the development of new innovative applications and to the development and operation of space-based infrastructures.

6. MAIN FINDINGS AND RECOMMENDATIONS

Figure 6.1. **Concept of the recommendations: "Building a sustainable bridge to the future for the governmental and private actors active in the space sector"**

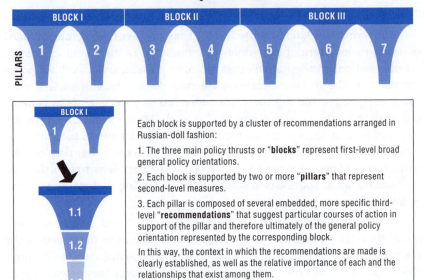

Each block is supported by a cluster of recommendations arranged in Russian-doll fashion:

1. The three main policy thrusts or "**blocks**" represent first-level broad general policy orientations.

2. Each block is supported by two or more "**pillars**" that represent second-level measures.

3. Each pillar is composed of several embedded, more specific third-level "**recommendations**" that suggest particular courses of action in support of the pillar and therefore ultimately of the general policy orientation represented by the corresponding block.

In this way, the context in which the recommendations are made is clearly established, as well as the relative importance of each and the relationships that exist among them.

Moreover, to improve legibility, a common format is adopted in order to present the recommendations in a consistent and systematic manner and link them to general policy principles. For each recommendation, the following structure is used:

- *General view* sets the context in which the recommendation is made and provides the general rationale for the recommendation.
- *Why* explains why government action is needed.
- *How* outlines actions governments might take to follow the recommendation.
- *Examples* illustrate how the proposed actions might be implemented in practice. The examples are included for illustrative purposes only.

The purpose of the framework is to outline what should be done on the basis of broad policy principles that reflect policy work conducted in the OECD in recent decades. As such, it can be used as a guide to a general assessment of the policies adopted in member countries. However, such an assessment is clearly beyond the scope of this report, although it might be a logical follow-up to the current project.

Because the approach is systematic, a number of the recommendations included in the framework are not new. They need to be included because they represent important elements of the overall architecture.

The term "special focus" identifies specific recommendations that may open interesting new perspective. Finally, throughout the text, issues or areas that may offer scope for further work for the OECD as a follow-up to this project are flagged.

Overview of the Recommendations

Block I – Implement a Sustainable Space Infrastructure

Pillar 1: *Implement a sustainable user-oriented space infrastructure*

- *Recommendation 1.1*: Foster the development of a more effective Earth observation infrastructure that allows for greater participation by both public and private actors
- *Recommendation 1.2*: Foster the development of an effective and sustainable satellite navigation infrastructure, fully suitable for public and commercial applications
- *Recommendation 1.3*: Encourage further development of communications satellite infrastructure suitable for meeting effectively both public and private needs.

Pillar 2: *Develop and maintain a cost-efficient space transport and servicing infrastructure*

- *Recommendation 2.1*: Encourage long-term R&D targeted to reducing the costs of access to space
- *Recommendation 2.2*: "*Special Focus*" Encourage international co-operation for conducting pre-competitive R&D work to reduce the cost of access to space
- *Recommendation 2.3*: Review "access to space" policy to reduce redundancy
- *Recommendation 2.4*: "*Special Focus*" Encourage long-term efforts to develop a sustainable in-orbit servicing infrastructure

Block II – Encourage Public Use

Pillar 3: *Encourage public use at national level*

- *Recommendation 3.1*: Create mechanisms for the effective generation and use of space-based data
- *Recommendation 3.2*: "*Special Focus*" Strengthen co-operation between user ministries and space agencies

Pillar 4: Encourage public use at international level

- *Recommendation 4.1:* Encourage use of space applications for global disaster prevention and emergency management purposes
- *Recommendation 4.2:* Encourage use of space applications for monitoring international treaties
- *Recommendation 4.3:* "*Special Focus*" Encourage the use of space applications to foster social and economic development in low-income countries

Block III – Encourage Private-sector Participation

Pillar 5: Create a supportive legal and regulatory environment for commercial activities

- *Recommendation 5.1:* Develop national space laws if they do not exist or complement existing ones
- *Recommendation 5.2:* Make existing space laws and regulations more business-friendly
- *Recommendation 5.3:* Adapt international space laws to business needs
- *Recommendation 5.4:* "*Special Focus*" Review the application of general laws and their impact on the development of space applications

Pillar 6: Strengthen private provision of space goods and services

- *Recommendation 6.1:* Foster public procurement from the private sector
- *Recommendation 6.2:* Privatise commercially viable business-oriented government activities
- *Recommendation 6.3:* "*Special Focus*" Encourage entrepreneurship and innovation

Pillar 7: Foster a more supportive international business and finance environment

- *Recommendation 7.1:* Extend free trade discipline to open up markets
- *Recommendation 7.2:* Encourage international standards
- *Recommendation 7.3:* Improve the allocation of spectrum and orbital positions
- *Recommendation 7.4:* Encourage the private financing of space activities

6. MAIN FINDINGS AND RECOMMENDATIONS

BLOCK I

Implement a Sustainable Space Infrastructure

The existence of an efficient, robust and sustainable infrastructure is essential for the smooth operation of modern economies. Major failures clearly demonstrate our dependence on the discrete, but ubiquitous, presence of such infrastructures as communication or electricity networks, as well as on their seamless and almost invisible but essential operation.

The same is true for space: without an efficient, robust and sustainable infrastructure, space actors will not be able to deliver, in an effective and efficient manner, space-based services that meet societal needs relating to major social challenges, such as those considered in this report. Hence, the development of a sustainable space infrastructure is viewed here as a major policy thrust that should be given particular attention by governments.

The term "space infrastructure" is defined as encompassing all space systems, whether public or private, that can be used to deliver space-based services. They include both space and ground segments.

This first block of recommendations is supported by two pillars:

- The first focuses on the "front office", *i.e.* that part of the space infrastructure that is "user-oriented" and designed to provide services to society at large; it currently includes communications, navigation and Earth observation services.
- The second addresses efforts that governments may wish to consider to strengthen the "back office", *i.e.* the space transport and servicing infrastructure. This is a critical enabling infrastructure that will play a major role in the development of the user-oriented infrastructure and, more generally, of the space sector in the coming decades.

The term "space transport and servicing infrastructure" is given here a broad interpretation. It encompasses not only the facilities needed to launch payloads into space and eventually bring them back to Earth, but also the support functions – such as in-orbit servicing and the management of space debris – that will be increasingly essential for the effective operation of the user-oriented space infrastructure over the coming decades.

Pillar 1: Implement a sustainable user-oriented space infrastructure

General view

Users will take advantage of space-based services only if: i) they truly meet user needs; ii) are cost-effective; and iii) are provided by an infrastructure that is robust, sustainable and fully integrated with ground-based facilities. This is because making use of a particular service generally requires substantial investment in terms of time and resources, notably in the case of space-based or space-enabled services for which the space component represents often only a very small, albeit essential, segment of the value chain. Entrepreneurs who are considering taking advantage of space-based solutions to serve their clients will be reluctant to undertake the necessary investment unless they have reasonable expectations that the space-based service they depend upon will continue to exist in a sustained, reliable and consistent manner in the future. Moreover, in some instances, the value of a service is directly linked to its duration, for example when the use of Earth observation data in climatology is an input for the understanding of long-term phenomena.

Why

Sustainable infrastructure cannot be developed without strong support from the state, either because the infrastructure is public or because the state provides a stable environment for the financing and operation of private operators.

When the private operator is in a situation of natural monopoly (because economies of scale are substantial and the efficient size of operation is large compared to the size of the market), an appropriate regulatory framework is needed to ensure that the service provided is sustainable and that the operator does not abuse its dominant position.

When several operators are in a position to compete, sustainability is less an issue for the state, as it becomes a competitive feature: successful operators will be those who offer users the most attractive combination of service quality, sustainability and price. For instance, in the competitive satellite communications market, sustainability is determined largely by the long-term contracts established between satellite operators and direct-to-home (DTH) platform operators. However, the state still plays a role by providing a stable regulatory framework and by ensuring systemic sustainability, *i.e.* ensuring that the failure of one or several private operators does not unduly affect the service provided to clients of failed operators and does not threaten the sustainability of the whole system.

How

- Foster the development of a more effective Earth observation infrastructure that allows for greater participation by both public and private actors (Recommendation 1.1).
- Foster the development of an effective and sustainable navigation system, fully suitable for civil public and commercial applications (Recommendation 1.2).
- Encourage further development of satellite communications infrastructure suitable for meeting effectively both public and private needs (Recommendation 1.3).

> **Recommendation 1.1: Foster the development of a more effective Earth observation infrastructure that allows for greater participation by both public and private actors**

General view

In the future, Earth observation (EO) systems are likely to play an increasingly important role in a growing range of activities. They will provide a unique capability – in close association with ground-based systems – to generate the data and information that will be needed to better understand, and better cope with, major problems to be faced in coming decades (ranging from problems raised by climate change and the management of natural resources to security challenges). EO systems will also play a significant role in providing emergency assistance worldwide and in helping low-income countries in their development efforts.

Why

Although Earth observation has substantial potential societal value, this potential is far from being fully exploited. One reason is that although EO systems are regional or even global in terms of coverage, decisions to deploy such systems are made at national level, in response to national objectives. These decisions can lead to co-operation for some applications (*e.g.* meteorological services), but they often led to duplication, as each space-faring nation seeks to use its own EO system to obtain independent access to similar information that it considers strategic. Moreover, there are gaps in coverage, not only because of resource limitations, but also because overarching global objectives are not always taken into account in the design of national systems.

Hence, given the public good nature of many of the services provided by Earth observation and given that the public sector is the main customer for many of these services, a co-operative public effort is needed at international level:

- To harmonise existing systems.
- To fill the gaps in coverage by developing new EO capabilities and launching new generations of EO satellites.
- To ensure the sustainability and interoperability of the various systems as well as the quality of the services they provide.
- To better integrate Earth observation capability with ground facilities.

How

- By identifying existing gaps and duplication, in close co-operation with users and in light of their data requirements.
- By estimating the potential benefits that might be achieved by upgrading the infrastructure as well as the costs involved.
- By establishing an overall plan in close co-operation with all stakeholders for the development of a sustainable system of systems.
- By establishing clear rules regarding the role to be played by private and public actors, including the role of governments in archiving and metadata harmonisation.
- By establishing appropriate institutional, funding and monitoring mechanisms for the deployment of space-based and ground-based upgrades.
- By implementing the plan in a co-ordinated manner.

Examples

- *Co-ordinating an international and sustainable system of systems: the GEO initiative*: As shown in the third phase of the project, efforts in this direction are already under way in the context of several international initiatives, including the Group on Earth Observations (GEO), which aims to move towards a comprehensive, co-ordinated and sustainable international EO system of systems. The future Global Earth Observation System of Systems should contribute to fuller integration of various observing systems, including space-based systems. GEO identified nine areas that will benefit from this international endeavour: natural and human-induced disasters; water resources; terrestrial, coastal and marine ecosystems; human health and well-being; energy resources; sustainable agriculture and desertification; weather information, forecasting and warning; climate variability and change; and biodiversity (see Chapter 5).

6. MAIN FINDINGS AND RECOMMENDATIONS

- *Global Monitoring for Environment and Security (GMES)*: this initiative, which is co-led by the European Commission and the European Space Agency (ESA), is a co-ordinated effort to provide integrated information services in support of the European Union (EU) and EU/ESA member state policies. Certain elements of GMES will provide the European contribution to the international GEO initiative (see Chapter 4).
- *Establishing a national dual-use system:* The Italian COSMO-SkyMed system (Constellation of Small Satellites for Mediterranean Basin Observation) developed by Alenia Spazio on behalf of the Italian Space Agency and the Italian Defence Ministry is a dual system for civil and intelligence/defence applications. This synthetic aperture radar system will allow observation in all visibility conditions (i.e. at night, through clouds) at high resolution, complementing other existing national/European systems in the framework of GMES. The first satellite is to be launched in 2005 (see Chapter 3).

Recommendation 1.2: Foster the development of an effective and sustainable satellite navigation infrastructure, fully suitable for public and commercial applications

General view

Like time-keeping, the ability to locate one's position or the position of various objects accurately and reliably is a fundamental and universal need in a modern economy, with wide-ranging implications for traffic management, security, the environment, the management of natural resources and the provision of personal services (civil and commercial). In fact, time-keeping and navigation are closely related since global navigation satellite systems (GNSS) are time-keeping devices with many uses, including navigation. Hence, navigation systems – whether terrestrial or space-based – provide an essential service that will be even more critical in an increasingly mobile society.

Why

Government action is needed because the substantial societal benefits to be derived from navigation systems will not be realised without public support. Moreover, governments have a clear interest in GNSS on strategic and security grounds.

As more and more activities come to depend on navigation systems, it is important that these systems meet standards that are as high as possible in terms of integrity, availability and accuracy. They also need to be sustainable so as to offer the continuity of service demanded by users. This applies to "safety-of-life" applications but also to commercial ones. Entrepreneurs seeking to develop applications are unlikely to invest substantial resources in such ventures unless the signal provider is in a position to guarantee a high degree of reliability and continuity of service.

How

- By pursuing efforts under way to develop Galileo as well civil and commercial applications that take advantage of the capabilities of the new system.
- By creating a permanent mechanism, *e.g.* by means of an international agreement or convention, preferably at global level, to ensure that existing and future GNSS (*e.g.* GPS III and an upgraded version of GLONASS) are developed and operated in a co-ordinated manner, including appropriate legal and regulatory mechanisms.
- By assessing how ground-based technology can be used to extend the use of space-based navigation systems (*e.g.* in urban areas, indoors).
- By pursuing ongoing work to develop public space-based and ground-based signal augmentation systems and encourage the private development of such systems when appropriate, in order to meet the needs of particularly important applications at regional and national levels.
- By maintaining an open regime for the production of navigation equipment and services worldwide.
- By establishing national and international legal regimes for the provision of commercial signals in which the obligations of the signal provider and its liability in case of failure are clearly defined.

Examples

- *Ensuring interoperability – the GPS/Galileo agreement:* The agreement reached in June 2004 between the United States and Contracting Parties to the Treaty establishing the European Community strengthens co-operation in the promotion, provision and use of the civil GPS and Galileo navigation and timing signals and services, value-added services, signal augmentation systems, and global navigation and timing goods. Notably, the decision that "GPS and Galileo shall be, to the greatest extent possible, interoperable at the non-military user level"[1] paves the way to a degree of harmonisation between the two systems that will provide civil users with higher signal integrity, accuracy and availability than either system individually. When Galileo becomes fully operational, the accuracy of the signal available to users throughout the world could rise with interoperability from around 20 metres 95% of the time to 1 metre 95% of the time. The co-ordinated use of the two infrastructures (double source) also strengthens security by reducing the adverse consequences of the failure of either. The agreement could facilitate the negotiation of a similar agreement with Russia regarding Galileo-GLONASS and GPS-GLONASS interoperability (see Chapter 4).

- *Creating a space-based augmented system:* The European Geostationary Navigation Overlay System (EGNOS) consists of responders on three geostationary satellites (two Inmarsat III satellites and ESA's Artemis satellite) and a network of ground stations for the transmission of a signal containing information on the reliability and accuracy of the positioning signals emitted by the American Global Positioning System (GPS) and the Russian GLONASS system. EGNOS will offer an accuracy of 5 metres (95%) in Europe. The EGNOS satellites that will be an integral part of the Galileo system are already in orbit. Early tests show even better accuracy than expected (1-2 metres 95% of the time, both horizontally and vertically). An important application of EGNOS will be active road management assisted by satellites, including virtual tolling, vehicle reporting (*e.g.* sending an emergency SOS that gives the exact position of the vehicle) and situational awareness (*e.g.* to warn motorists of hazards ahead) (see Chapter 4).

- *Promoting international co-operation in satellite navigation – The Galileo Joint Undertaking:* As a policy move to promote international co-operation in satellite navigation and share some of the development costs, the European Commission and the European Space Agency have welcomed equity participation in the Galileo system by non-European countries. For instance, China is expected to participate in the development of some systems. Other possible candidates include Israel, India, Korea, Brazil and Mexico (see Chapter 4).

> **Recommendation 1.3: Encourage further development of communications satellite infrastructure suitable for meeting effectively both public and private needs**

General view

Satellite communication is an essential element of the communications infrastructure. Satellites have specific advantages in terms of international coverage, broadcasting, flexibility and rapid deployment of service. They have been very successful in some market segments (*e.g.* DTH services) and help to foster competition and innovation in those markets. They also provide the technical means for delivering some public services (*e.g.* for serving rural and remote areas, for delivering emergency services).

Why

It is essentially the private sector that develops the civil satellite communications infrastructure. However, governments play an important supporting role to ensure that the communication infrastructure and all its main components evolve as needed to fully support economic and social development and the move to a knowledge society. The communications infrastructure should also provide an effective tool for delivering public services (such as health and education) to all in a cost-effective manner, including to households in rural and remote areas. Governments also have a responsibility to ensure that technology-biased regulations do not bar telecommunications satellites from providing services in competition with terrestrial solutions.

How

- By fostering a more level playing field for satellite operators. In a number of markets, satellite operators face significant barriers to entry, notably when entrenched incumbents – with significant sunk costs in their existing infrastructure – dominate such markets.
- By providing R&D support for the development of new innovative space-based communication technologies. The development of a new generation of telecommunications satellites calls for high-risk R&D investments that can generate significant externalities for society at large.
- By encouraging their deployment when it is the most cost-effective way to extend the delivery of e-government services to all citizens. This applies notably to the provision of such services in rural and remote areas so as to reduce the digital divide. Given the large economies of scale that prevail in the provision of satellite-based services, this may call in some cases for measures designed to facilitate the aggregation of demand across large geographical areas.
- By encouraging the development of open industry standards that help to cut costs and foster the scalability of systems while maintaining incentives to innovate.

Examples

- *Two high-value societal applications – telehealth and distance learning:* As the third phase of the project made clear, telehealth services are likely to expand significantly in the coming decades as the demand for efficient health services in the home or in the local community increases and as several converging technologies greatly enhance and facilitate the electronic delivery of such services. Space solutions are attractive, notably for patients living in rural and remote areas, for patients on the move or for

6. MAIN FINDINGS AND RECOMMENDATIONS

accident victims. Satellites also provide an ideal tool for distributing large amounts of educational and training material over large areas, notably in the developing world. Satellite-based video conferencing services can also be used for keeping professionals (*e.g.* doctors) up to date. India will exploit this capability on a large scale with the recent launching of EduSat, the first satellite fully dedicated to education (see Chapter 2).

- *Providing R&D support – The Agora programme:* France's CNES launched in the summer of 2004, in co-operation with industry, an R&D programme dubbed Agora (*Accès garanti et optimisé pour les régions et l'aménagement du territoire*) to develop a new generation of spot-beam satellites, offering three broadband telecommunications services at competitive costs: high-speed Internet access, interactive television programmes and Voice-over-IP. A first geostationary satellite could be launched in 2007 and a second in 2009. Although CNES is the focal point at present, Agora should become a privately owned infrastructure, managed and operated by private operators (see Chapter 4).
- *Providing public services in rural areas:* Communications satellites can be used to provide public services in rural areas, as shown by the French postal service (*La Poste*) which recently installed satellite-based *kiosques* in a number of French villages, giving local inhabitants access to postal services, the Internet and e-mail services (see Chapter 4).

> **OECD follow-up:** Explore further the costs and benefits provided by space systems.

Pillar 2: Develop and maintain a cost-efficient space transport and servicing infrastructure

General view

An efficient and cost-effective space transport infrastructure is essential for the development and proper maintenance of an effective and sustainable service-oriented space infrastructure. This is far from the case today:

- Access to space and the operation of space assets remain very risky and costly. Problems are likely to increase as the volume of space debris expands.
- It is currently technically impossible or far too expensive, in most cases, to service space assets once they have been launched. This contributes to system obsolescence and to the need for operators to launch costly back-up satellites.

- The private component of the space transport industry is not in good financial shape. It faces a situation of chronic excess supply, owing to the desire of space-faring nations for independent and assured access to space.
- Markets for launch services are largely protected, inhibiting the development of a competitive market and reducing incentives for industry to spend money on becoming more efficient.

Why

Governments have the main responsibility for the situation currently faced by the space transport industry and are best placed to take corrective action. First, they are the logical source of an appropriate legal and regulatory framework that encourages innovation and a more efficient use of resources. They also have the main responsibility for undertaking the basic R&D that will be needed to reduce the cost of access to space and to develop the enabling technologies needed to deploy a truly effective in-orbit service facility. Finally, governments have a collective responsibility to prevent the further cluttering of space with harmful debris and for taking measures to gradually eliminate such debris from outer space. The private sector can play an important supporting role in developing new systems and new applications when it is cost-effective to do so.

How

- Encourage generally long-term R&D targeted to reducing the cost of access to space (Recommendation 2.1).
- Encourage international co-operation for conducting pre-competitive R&D work to develop generic or enabling technologies (Recommendation 2.2).
- Review "access to space" policy with a view to alleviating – if possible – the chronic excess supply conditions that prevail in the launching and launcher manufacturing segment of the industry (Recommendation 2.3).
- Encourage long-term efforts to develop a sustainable in-orbit servicing infrastructure and the establishment of an international debris regime for controlling, monitoring and possibly mitigating man-made space debris (Recommendation 2.4).

Recommendation 2.1: Encourage long-term R&D targeted to reducing the cost of access to space

General view

Sending payloads into space is a formidable challenge. It is very costly and very risky. The only technology that can be used effectively today (expandable launch vehicles – ELVS) has experienced only incremental improvements over the last 50 years. A paradigm shift, or technology breakthrough, is needed to achieve significant progress.

In future, the main challenge will be to develop new technologies that can effectively help to cut costs and improve the reliability of launchers significantly, *i.e.* by several orders of magnitude. This includes progress in technologies such as space propulsion, space transport systems and orbital systems. Non-launch vehicle costs account for about one-third of the total cost of access to space. Hence, a reduction in the cost of various launching services (*e.g.* safety costs, launch site facilities costs) is also important.

Why

Reducing the cost of access to space is a quasi-public good, since all users of space services would benefit, including the public users that are the largest customers of launching services. Hence, governments have a direct interest and a decline in space access costs would also allow space agencies to do more with existing budgets.

Governments are also best placed to take on the risks and sustain the research effort over the sufficiently long period of time that would be needed to achieve results, in co-operation with private actors (including small innovative entrepreneurs keen to develop new markets). However, priority for such longer-term R&D efforts has not always been maintained in the past, as long-term programmes are often the most vulnerable to budget cuts. Indeed, many of the world's launch programmes have a history of being cancelled only a few years after being started. One reason is that strategic objectives (*e.g.* achieving independent access to space) have generally prevailed over long-term cost considerations and induced the main space actors to give the preference to mature technologies that can more quickly meet the strategic objective.

How

- By encouraging space agencies to devote a significant share of their budget to basic R&D work on a sustained basis in order to address effectively over time the major technological challenges that will need to be overcome.

- By encouraging industry to participate in such efforts, including in particular space entrepreneurs with no vested interest in the existing transport infrastructure. This could include the setting up of prizes (*e.g.* the X-Prize), tax reductions[2] and liability-sharing regimes (*e.g.* maximum probable loss regime in US and Australian national space laws).[3]
- By encouraging inter-agency work between space and non-space organisations, to provide diversified support to sustain some R&D projects over the long term or simply to benefit from research conducted elsewhere.

Examples

- *Development of demonstrators – the X-43A approach:* Over the past 40 years, the United States has spent over USD 4 billion on hypersonic technologies with a view to developing a reusable first stage. This is a very modest amount when compared to the budget devoted to the shuttle and the International Space Station (ISS). NASA spends more on the shuttle in a single year (USD 5 billion in 2005, USD 4.3 billion expected in 2006). Moreover, not only have the budgets allocated to hypersonic technologies been modest, they have also suffered from a lack of long-term commitment. One of the most recent such efforts is the X-43A, NASA's unpiloted and non-recoverable aircraft that demonstrates air-breathing scramjet technologies, as part of NASA's Hyper-X programme. In March 2004, it made its second experimental flight and reached Mach 7 over the Pacific Ocean. This was far faster than any air-breathing aircraft has yet flown (the world's fastest air-breathing aircraft, the SR-71, cruises slightly above Mach 3). In November 2004, a third X-43A demonstrator broke a new record, demonstrating that an air-breathing engine can fly at nearly ten times the speed of sound, as it successfully cruised at nearly Mach 9.8 for a few seconds (see Chapter 3).
- *Development of demonstrators – the Phoenix:* In Europe, more limited R&D work is conducted on renewable launch vehicles (RLVs). The Phoenix is the prototype of a future RLV with 50% financing by European industry (EADS Space Transportation and OHB System of Bremen, Germany). The Phoenix has also received funding from the German federal and Bremen state governments. Total financing is about EUR 16 million (USD 17.4 million). The first fully automatic test flight was carried out successfully in May 2004 (the demonstrator was dropped from an altitude of 2 400 metres by a helicopter). Further developments of the Phoenix could be incorporated into an ESA programme to study possible follow-ups to the Ariane launcher programme. However, ESA's planned multi-year Future Launcher Preparatory Programme has been postponed several times over the past years owing to the financial cost of putting Ariane 5 back on track (see Chapter 3).

- *Following up the X-Prize win:* The X-Prize competition, won in October 2004 by the SpaceShipOne team, has had an impact on R&D in the private sector, but also in space agencies. SpaceDev, the company that designed and built the hybrid rocket engine for Scaled Composite's SpaceShipOne, announced in September 2004 that it had been awarded a contract by NASA to design a low-cost suborbital spacecraft, the Dream Chaser. The Dream Chaser would take off vertically, and carry up to three people to an altitude of 160 km (100 miles). The spacecraft could be built by 2008, and would demonstrate a set of launch and flight technologies. Further versions of the spacecraft could eventually go into orbit and compete with other proposed spacecraft to transfer crew to and from the International Space Station (see Chapter 4).

- *Space and non-space organisations joining forces for R&D – NASA and the Department of Energy:* Because NASA has to respect the openness mandated by the National Aeronautical and Space Act of 1958, NASA employees must make public information they have acquired. However, this does not prevent the agency from working on secret projects with other government agencies: for instance, specific arrangements with the Department of Energy regarding research on space nuclear power (project Prometheus) allow NASA to respect its mandate, while Department of Energy keeps a part of their work undisclosed (see Chapter 4).

> **Recommendation 2.2: "*Special Focus*" Encourage international co-operation for conducting pre-competitive R&D work to reduce the cost of access to space**

General view

"Pre-competitive R&D" refers to R&D that is distant from the market and focused on "generic" or "enabling" technologies rather than on technologies targeted at particular markets. Typically, the research effort is not expected to produce commercially usable technologies or products but to reach the stage of demonstrating feasibility or providing research prototypes.

Co-operation at the pre-competitive stage is often a good way to overcome basic technological hurdles. Such collaborative efforts have been undertaken extensively over the last decade in many OECD countries, often at national level. One example is the US SEMATECH (SEmiconductor MAnufacturing TECHnology) programme, which was developed for national security reasons to counter a loss of technological capability to foreign competitors. Such efforts can, however, be undertaken at international level if the parties share a common concern and all can benefit from the work. For instance, pre-competitive research is an important feature of the EU framework programmes for research and technological development. The sixth of these programmes identified space as a priority field and the seventh perhaps even more so.

Why

Reducing the cost of access to space is a public good that would be beneficial for all nations. It is a common goal that would be more easily achieved if all major space-faring nations co-operated. The required effort is primarily at the level of generic and enabling technologies (*e.g.* propulsion) and is therefore well-suited to pre-competitive basic research.

How

- By encouraging the setting up of an international consortium in which space agencies and industry would share development costs. It would also provide a forum for addressing common technical challenges and setting clear technical benchmarks.
- By adopting special measures to deal with security concerns. For instance, a "black box" approach could be used to deal with strategically sensitive items.[4]

Examples

- *Development of multilateral agreements:* The European Space Agency and Russia have signed an agreement to co-operate on research and joint development of new space transport technologies. At industry level, some companies are also already starting to co-operate internationally on space and defence systems (*e.g.* transatlantic co-operation on anti-missile defence between Lockheed Martin and EADS) (see Chapter 3).
- *International collaboration on developing enabling technologies: the US-Japanese MB-XX engine:* The Boeing company's Rocketdyne Propulsion and Power unit in the United States and Mitsubishi Heavy Industries in Japan have worked together since 1999 on the design and development of the MB-XX engine, a new liquid oxygen/liquid hydrogen upper-stage engine for the next generation of expendable satellite launch vehicles. In 2002, they successfully completed the preliminary MB-XX full-scale combustion chamber/injector assembly test programme and plan to test the engine jointly in 2005. This co-operation on sensitive enabling space technology was only possible by working with "black boxes", each company keeping some elements from the other to respect in particular US technology transfer regulations (see Chapter 3).

> **Recommendation 2.3: Review "access to space" policy to reduce redundancy**

General view

The access to space policies adopted by most space-faring countries have typically three main elements: i) the development of one or more launchers at national (or regional level) with strong R&D support by government to ensure independent access to space; ii) preferential use of these launchers by institutional users; iii) efforts to offer the launcher on a commercial basis in order to recoup some of the development and fixed costs.

The overarching sovereignty and security concerns that drive this policy approach should logically take precedence over economic considerations. However, the cost that such a policy imposes on society is worth taking into account. In fact, the policy has two perverse effects from an economic perspective:

- It prevents the exit of firms that would fail under normal business conditions, thereby preventing an efficiency-enhancing reallocation of resources.
- It inhibits competition, since the firms' survival is achieved essentially by protecting their public market.

Moreover, this policy often tends to be self-defeating in practice, as attempts to recoup costs on the commercial market drive down the price of commercial launches. The net result is that launch prices do not always cover marginal costs and increase further the total cost of the systems.

Why

Because of the policy approach outlined above, the upstream segment of the sector (manufacturers of launchers and spacecraft) faces a costly situation of chronic oversupply with very short production runs. Short production runs, in turn, may have an adverse effect on the reliability of the launchers (hence their ability to provide effective access to space), as "learning by doing" plays a key role in this activity. It is up to governments to find collectively a better solution to the problem. The main policy challenge is to find ways to reduce the inefficiencies of the system, while fully taking into account legitimate security and sovereignty constraints.

How

- By reviewing their access to space policy, notably by considering whether the benefits – in terms of independent (or assured) access to space – are really worth the cost of maintaining artificially, among allies, an excess number of separate launchers.

- By exploring the possibilities of further opening the institutional markets to competition (i.e. scientific and other governmental payloads) among allies (*e.g.* public procurement issues).

- By exploring the possibility of extending the use of mutual backup agreements between launch providers (see example below) to institutional payloads.

Examples

- *Reviewing access to space policy:* Going beyond achieving "independent access" to space, the US Department of Defense (DoD) sought in the 1990s to develop a "guaranteed access" strategy. For this purpose, it adopted the evolved expendable launch vehicle (EELV) programme, which called for the production of two separate rockets (Atlas V and Delta IV), which would share virtually no common systems and be built on separate production lines, so as to ensure that the United States would have at all times at least one operational family of launch vehicles available. For 2006, the DoD requested USD 864 million for the EELV programme. Faced with rising cost overruns, the DoD announced that it intends to reassess its dual supplier policy for the EELVs programme by 2009 (see Chapter 4).

- *Mutual backup agreement between launch providers:* Currently, mutual backup agreements exist between Proton and Atlas vehicles (both provided by International Launch Services). More recently, Arianespace and Sea Launch LLC have joined with Mitsubishi Heavy Industries to offer mutual backup for the Ariane 5, Sea Launch and H-2A rockets. Customers have welcomed this development because it allows satellite owners to arrange launches and have greater confidence in the schedule (see Chapter 3).

> **Recommendation 2.4: "Special Focus" Encourage long-term efforts to develop a sustainable in-orbit servicing infrastructure**

General view

To run space infrastructure effectively and in a sustainable manner, operators should ideally be able to perform the servicing and maintenance of space platforms (*e.g.* satellite, space station) efficiently on a routine basis. This would involve, for example, the ability to replenish consumables and degradables (*e.g.* propellants, batteries, solar array); to replace failed functionality (*e.g.* payload and bus electronics, mechanical components); and to enhance the mission (*e.g.* software and hardware upgrades). Such servicing should also logically encompass the removal of debris and the orderly disposal of satellites at the end of their useful lives.

6. MAIN FINDINGS AND RECOMMENDATIONS

Until now, in-orbit servicing has been limited to human missions (*e.g.* shuttle missions to repair the Hubble telescope) and software upgrades (*e.g.* scientific missions). The main limitation is cost and the fact that spacecraft typically are not designed with servicing in mind. At the same time, no progress has been made regarding the cleaning up of space debris. Indeed, as more and more assets have been put in orbit, the amount of man-made space debris has increased substantially, both in low Earth orbit and along the geostationary arc 36 000 kilometres above the equator.

The analysis carried out in the second phase of this project suggests that, over the longer run, the demand for in-orbit servicing is likely to increase in a broad range of possible futures. Indeed, the ability to service spacecraft would enable operators to provide more reliable service with less need for expensive back-up satellites and would allow them to keep their spacecrafts' electronics up to date. The weight and cost of satellites would also be reduced if refuelling is easier and cheaper. There will also be a growing demand for dealing more effectively with the debris problem. If this problem is not effectively tackled in the coming years, it could ultimately shut down the space industry or at least require the use of expensive "hardening" technologies.

Why

In-orbit servicing has a strongly positive public good value: once developed, everyone will benefit. Hence it is clearly the responsibility of governments to encourage the development of the basic technology. By contrast, space debris has a strongly negative public good value. It is a highly undesirable and potentially very harmful form of space pollution. Hence, governments have a collective responsibility to create, at international level, an appropriate legal and regulatory framework for ensuring the mitigation and possible future disposal of space debris.

How

- By encouraging space agencies, in co-operation with satellite operators and manufacturers, to research and develop a new generation of serviceable satellites and other platforms.
- By encouraging R&D efforts towards progress in robotics and other technologies for performing servicing functions (*e.g.* docking with the satellite, plugging new fuel cell and hardware modules).
- By ensuring that a proper legal framework is developed for these new activities (*e.g.* in terms of liability and insurance obligations, salvage rules, frequencies necessary, environmental and disposal issues).
- By establishing an international legal and regulatory regime, on the basis of the recommendation made by the Inter-Agency Space Debris Co-ordination

Committee (IADC), with appropriate enforcement mechanisms to implement debris mitigation regulations at national level, based on a mutually agreed model regime. Non-compliance could be penalised by the imposition of fines or barring the offender from offering his products and services in some markets.[5]

Examples

- *Public in-orbit servicing efforts – the first steps:* Although technical constraints are important, some technologies and mission operations already developed for specific orbital purposes pave the way towards the development of a fully fledged in-orbit servicing facility for high-value satellites. Examples include the robotic solution to service Hubble, Europe's Automated Transfer Vehicle (ATV) and its automatic docking technology to the International Space Station) (see Chapter 3).

- *A commercial approach – Orbital Recovery Corporation (ORC):* This company has in the development stage an orbital space tug called the Orbital Life Extension Vehicle (OLEV), whose purpose is to mate mechanically with an existing communications spacecraft in the 36 000 km geostationary orbit (GEO) or GEO-intended orbit, with a sufficient amount of fuel to keep the satellite in GEO for up to an additional ten years of life. The OLEV could also be placed in orbit above or below the GEO belt so as to respond rapidly to failures of upper stages or the in-orbit propulsion systems of GEO assets. The OLEV can also offer de-orbiting services (see Chapter 3).

- *The US Federal Communications Commission's regulations on space debris:* The Federal Communications Commission (FCC) has recently issued regulations for communication satellites regarding end-of-life orbits and space debris. The regulations provide that all US-licensed satellites launched after 18 March 2002 have to be placed into so-called graveyard orbits between 200 km and 300 km above the geostationary arc, where most commercial satellites operate. This could set a regulatory standard that could be adopted by other nations and lead eventually to a broad-based international agreement (see Chapter 5).

> **Special Focus:** This recommendation calls for special attention from decision makers, as R&D on in-orbit servicing and its related enabling technologies (*e.g.* robotics) might provide the basis for technological breakthroughs in space and non-space sectors. By developing new orbital techniques and operations in co-operation with the private sector, a new paradigm for the overall space infrastructure would be tested, with potential long-term benefits for the sector and for users.

Notes

1. Agreement on the Promotion, Provision and Use of Galileo and GPS Satellite-Based Navigation Systems and Related Applications, Article 4.3.

2. Though they are generally helpful, tax reductions and other fiscal incentives still have the most beneficial effects when there is already a clear revenue stream from a project's operations, rather than in the early R&D phases.

3. Under Australian space law, the launch operator must have insurance covering the maximum probable loss. If a foreign country, on its own behalf or on behalf of a citizen, makes a claim against the Australian government, the launch operator is liable to indemnify the government only up to the insured amount based on the maximum probable loss.

4. The legal aspects of such mechanisms, *e.g.* with respect to liability claims, have to be dealt with through appropriate legal arrangements, at either national or (preferably) international level.

5. The guidelines promulgated by the IADC would provide an excellent point of departure for establishment of such an international agreement.

BLOCK II

Encourage Public Use

Typically, governments are major users of infrastructure, whether they use public infrastructure to deliver public services to citizens or whether they use the services of private infrastructure as an input in their activities. In most cases, public services on offer are financed by general taxes on the population at large and are provided free of charge or on the basis of marginal cost of provision.

Space infrastructures offer very attractive opportunities to pursue a broad range of public missions in a cost-effective manner. The utilisation of space assets can help address long-term societal needs such as those related to the environment, the management of natural resources, security, mobility and the move to a knowledge society. Unfortunately, such opportunities are not always fully exploited for a variety of reasons, ranging from lack of information to regulatory constraints or the existence of rigid bureaucratic rules that prevent the effective use of the infrastructures. Hence, a systematic approach that takes full account of all major impediments is needed to encourage the use of space infrastructure when it is cost-effective to do so.

Two pillars support this block of recommendations:

- **Pillar 3** focuses on public initiatives for encouraging public use at national level. The main thrust is on measures designed to strengthen co-operation between user ministries and space agencies in order to foster the effective generation and use of space-based data and facilitate transactions between suppliers and public users of space-based services.
- **Pillar 4** addresses government actions at international level that make it possible to take better advantage of the ubiquity that space-based services offer. These recommendations include a broad grouping of public initiatives, ranging from risk prevention, to emergency support and assistance to developing countries in the management of their resources, to the monitoring of the effective application of international treaties.

Pillar 3: Encourage public use at national level

General view

Space solutions can help governments fulfil their mission in a number of areas, ranging from environmental policy to transport to civil protection. For instance, public agencies responsible for the delivery of education and health services can take advantage of satellite communications to extend the reach of their services. As well, public agencies responsible for agriculture, the management of natural resources and territorial development can exploit available Earth observation data to enhance their information management systems and improve their ability to make effective and timely decisions. Another important area of application involves transport agencies and ministries that can use global navigation satellite systems to improve the movement of persons and goods and reduce the social costs imposed by pollution, congestion and accidents.

Why

Governments have the main responsibility for using the existing space infrastructure as efficiently as possible in the pursuit of public objectives. However, the potential that space can offer has not been fully exploited:

- The data and information provided by the Earth observation infrastructure is not always used effectively.
- The potential of satellite communications for extending e-government to rural and remote areas is not fully exploited.
- The use of navigation infrastructure is still at an early stage; further technical and cost/price incentives are needed to apply it to traffic management, emergency services and other transport-related uses.

How

- Create mechanisms for the effective generation and use of space-based data (Recommendation 3.1)
- Strengthen co-operation between user ministries and space agencies (Recommendation 3.2).

> **Recommendation 3.1: Create mechanisms for the effective generation and use of space-based data**

General view

Effective generation and use of data and information is a major prerequisite for government action. Without timely and pertinent data, decision makers are not alerted early enough to new problems to have the time and information necessary to take effective corrective action expeditiously, to monitor progress closely and to verify thoroughly that the results achieved actually conform to expectations.

In some cases, the development of appropriate databases does not raise any particular issues, either because the producers of the data are also the ones who use them, or because data users and producers work closely together. In such cases, data producers fully understand what users need and users know precisely what producers can offer. In other instances, however, things are more complicated: data are from various sources and use different, independently developed formats that are often incompatible, while the user community is diverse and fragmented. In such cases, considerable inefficiencies arise: data are generated but not used because they do not meet users' requirements or are not in an appropriate format, or users do not have access to the data they need. A related problem is that even when the data are suitable, there are not enough information systems for exploiting data to meet users' needs. These shortcomings call for the development and implementation of a comprehensive government-wide data policy designed to promote better co-ordination of the actions of data producers and data users, as well as to reduce their transaction costs. Moreover, the development of information systems for exploiting such data should be encouraged.

Why

Governments have the main responsibility for the effective generation and use of Earth observation data. They have already devoted considerable resources to generating such data. It is in their best interests to develop a comprehensive data policy so as to ensure that quality data are generated and made available to users, that such data effectively meet users' needs, are easily accessible, are used as effectively as possible and are properly archived.

How

The establishment of a comprehensive data policy calls for the implementation of a number of complementary measures, including:

- *Measures to ensure control of data and information:* This is essential for assuring the quality of the data and protecting the rights of both data producers and users. Without proper control, the data producer cannot guarantee quality. This includes both physical control (*e.g.* encryption) and legal control (*e.g.* copyright and licensing agreements that document the rights and obligations of suppliers and users, and help to protect the quality of products and foster the recognition of the data in the community of users).
- *Measures to foster technical accessibility:* This requires the adoption of metadata standards. Without the use of agreed standards for the provision of data and products, the ability of users to take advantage of the potential benefits to be achieved by using the data is severely impaired. Also important would be measures to improve data storage and retrieval and to encourage the development of user-friendly information systems.
- *Measures to balance open access objectives and security concerns:* All data sources should be tapped, including dual use and military assets, and the information should be made available as widely as possible. However, measures need to be taken to protect data considered as "sensitive" from a security perspective.
- *Measures that promote effective funding of publicly generated data:* Adequate and sustainable sources of funding should be available for publicly generated data. In this regard, customer funding (i.e. by user departments) may be the best way to achieve sustainable funding and ensure that the data produced actually meet the requirements of users.
- *Measures to promote an effective pricing policy for private data providers:* Private data providers should be able to generate sufficient revenue to justify their investment. In this regard, clear rules should be established regarding how the "non-discriminatory access principle" and security concerns may apply to their activities. This would include a consideration of whether the UN Principles on Remote Sensing should be updated and/or renegotiated as a full treaty. As well, non-competition agreements between government and industry could be adopted.
- *Measures to facilitate transactions between data supplies and users:* When public demand is fragmented across a large number of user agencies, it can make sense to aggregate demand so that suppliers deal with a single agency. This also provides an opportunity for both parties to agree on licensing arrangements that allow broader use of the data.
- *Measures to ensure proper archiving of the data:* When various public and private producers generate data, data archiving is not always consistent and systematic. Hence, a mechanism is needed to ensure the preservation of

data. For this purpose, an appropriate legal instrument should give a public body the right to accept data sets that are scheduled to be destroyed and sufficient funding to manage and preserve the data effectively, since national archives are generally financed and run by public authorities.

Examples

- *Facilitating transactions between data suppliers and users – ClearView:* In order to facilitate the relationship between data users and data producers, some governments have found it useful to aggregate demand for such data. For instance, the United States has a programme called ClearView for this purpose which is run by the US National Geospatial Intelligence Agency (NGA), which has replaced the National Imagery and Mapping Agency (NIMA). ClearView was set up in January 2003 to help the Pentagon procure commercial satellite imagery. It has a five-year contract with a base performance period of three years and two additional one-year options. The programme will replace what NIMA officials considered a cumbersome licensing structure with a single licence that allows imagery to be shared with all the agency's potential partners. The contracts established between the agency and private data providers give the agency and its customers access to commercial imagery that is extremely timely. Three providers of commercial imagery currently participate in the programme: Space Imaging, Digital Globe and Orbimage (see Chapter 4).

- *Developing a structured framework for data integration and information management – the Global Monitoring for Environment and Security (GMES):* The overall aim of the GMES initiative is to support the achievement of Europe's goal regarding sustainable development and global governance by providing a structured framework for data integration and information management so that users are provided with timely and quality data, information and knowledge. The system is expected to be fully operational before the end of the decade (see Chapter 4).

- *Establishing appropriate institutional, funding and monitoring mechanisms – the EUMETSAT model:* The European Organisation for the Exploitation of Meteorological Satellites (EUMETSAT) is an intergovernmental organisation with 18 European member states and nine co-operating states. Since 1995, EUMETSAT has had direct responsibility for the operation of its satellites in orbit and new programmes to ensure the continuity of observations. In particular, it is responsible: i) for the launch and operation of the meteorological satellites; ii) for delivering satellite data to end users (e.g. national meteorological services of member states and others); and iii) for contributing to the operational monitoring of climate and the detection of global climate change in international co-operation projects (see Chapter 4).

> **OECD follow-up:** In its regular analytical activities, the OECD deals with various sets of economic data and statistics from many countries. The Organisation's expertise in collecting and managing data – while verifying their validity and compatibility – could be shared with space-related entities.

Recommendation 3.2: "*Special Focus*" Strengthen co-operation between user ministries and space agencies

General view

Governments have wide-ranging obligations that force them to create an ever-increasing number of specialised agencies in order to deal more effectively with specific problems. Such agencies are typically organised in a hierarchical manner so that each ultimately reports to, and receives instructions from the central government.

A major drawback of this architecture is that communication and shared funding across public bodies is difficult. It is particularly serious when the specialised agency has a high level of expertise that is not available elsewhere in government (*e.g.* space agencies). In such cases, information asymmetry may hinder communication between the specialised agency and user ministries: experts in the specialised agency have little knowledge of users' needs, while users are not aware of how space can be applied in their domain of activity.

Why

The problem noted above is particularly serious for space, since the technology requires specialised expertise while its applications are wide-ranging. Space technology is an enabling technology: in many applications the space segment is often a relatively small – albeit essential – component of the value chain.

Potential users are often little inclined to learn how the information they need is actually produced. They are more concerned about the timeliness, accuracy and pertinence of the information and services.

Moreover, even if user agencies are aware of what space can offer, they may not have budgets earmarked for dedicated programmes (*e.g.* to buy maps with processed space data), and depend on space-related agencies for funding. However, space agencies are first and foremost research organisations and cannot be expected to support applications financially beyond the demonstration stage.

How

- By establishing formal co-operative mechanisms between users and producers. Such mechanisms (including internal regulations) facilitate the development of an ongoing dialogue between users and producers. This allows space agencies to become aware at an early stage of the needs of user departments, while user departments achieve a better understanding of what space can offer them. The sharing of experience between user department and space agencies offers opportunities for synergies and provides feedback that can be usefully applied to develop best practices. It also put users in a better position to take advantage of existing commercial space products and services. Space agencies also need to maintain regular contacts with associations of users, industry associations and industry actors.
- By establishing in each main user department a promoter – at a level high enough to be effective – who can increase the department's awareness of the benefits of satellite services.
- By encouraging key user departments to articulate their requirements and engage more strategically in the development of space services.
- By prioritising and then establishing significant demonstrator projects with a meaningful follow-on.
- By ensuring that appropriate financing mechanisms are in place to enable user departments to take full advantage of opportunities that space may offer for the effective delivery of public services.

Examples

- *Establish formal co-operative mechanisms between users and producers – the BNSC partnership:* This is a voluntary partnership of departments, agencies and research councils hosted by the UK Department of Trade and Industry (DTI) which co-ordinates policy and programmes. The partnership works closely with all stakeholders (scientific community, industry and other government users of space services) and seeks to develop synergies among existing interests, to ensure that space activities are co-ordinated so as to avoid duplication or gaps, and to deliver maximum benefits. A major added value is the BNSC's role in reaching out, with a long-term strategic view, to other areas of government where space services may be relevant for implementing public policy. The BNSC engages in dialogue with agencies that are not formally members of the partnership but have a growing interest in space (see Chapter 4).
- *Co-operation in fire management:* Several space-related agencies and industry have teamed up with national civil protection agencies in applying space applications to fire fighting operations (*e.g.* France's Centre National

d'Études Spatiales [CNES] and the French civil protection; the United States' National Oceanic and Atmospheric Administration [NOAA], US Geological Survey [USGS], Federal Emergency Management Agency [FEMA] and National Interagency Fire Center, as part of the US Hazard Programme). Increased links between developers of space applications and final users provide important "lessons learned" for responding to operational users' needs and requirements (see Chapter 2).

> **Special Focus:** This recommendation calls for special attention from decision makers. While a level of co-operation between space agencies and potential public users of space data and services exists in several countries, such co-operation is not always effective and does not always lead to meaningful follow-ons.

> **OECD follow-up:** The OECD has specific competencies for analysing institutional mechanisms in OECD and non-OECD countries. It could evaluate co-operation between space agencies and user ministries in several countries, with a view to comparing and identifying best practices, taking national particularities into account.

Pillar 4: Encourage public use at international level

General view

Owing to advances in communications and transport, policy makers are more and more required to respond to requests originating outside their borders. For instance, progress in communications means that people are better informed and made aware earlier of the occurrence of disasters in other parts of the world, while affected countries are in a better position to communicate their relief needs. At the same time, because of progress in transport, countries providing assistance are better able to act on such information in a timely manner. As a result, the scope for dealing with disasters at the international level is increasing.

Also, the growing mobility of goods across borders means that there is an increasing need to track such goods effectively. While this can be done at national level with existing terrestrial technologies, there is no overarching system in place to perform this function systematically from origin to destination, whatever the origin and destination.

At the same time, human action is causing externalities that are not confined to national borders. This includes, for instance, cross-border pollution, the generation of greenhouse gases, the depletion of fish stocks. Such problems can only be dealt with effectively at the international level.

Space systems can be an effective part of the solution to some of these problems because of their ubiquity, the non-intrusive nature of the services they offer and the fact that they can be rapidly deployed to theatres where their services are needed most, anywhere in the world.

Why

First, many of the needed actions fall in the public domain. Moreover, most of the space systems to be used for international missions are public or can be used by public authorities for this purpose.

Second, governments have a moral – and at least in part a legal – obligation to act as "good international citizens". This includes:

- Providing assistance to other nations in case of disaster.
- Helping to foster international relations and to address problems of an international nature.
- Providing assistance to developing countries and helping alleviate poverty and substandard living conditions.

How

- Encourage the use of space applications for global disaster prevention and emergency management purposes (Recommendation 4.1).
- Encourage the use of space applications for monitoring the movement of hazardous goods (Recommendation 4.2).
- Encourage the use of space applications for monitoring international treaties (Recommendation 4.3).
- Encourage the use of space applications to achieve social and economic development in low-income countries (Recommendation 4.4).

> **Recommendation 4.1: Encourage the use of space applications for global disaster prevention and emergency management purposes**

General view

The intensification of weather extremes, natural and technological hazards and the resulting increase in potential economic losses present new challenges for decisions makers, emergency agencies and the insurance sector.

Where risk and disaster management at international level is becoming a main concern, space-based systems can provide specific capabilities for addressing those challenges, but some sustainability issues still need to be resolved.

Why

Space-based systems are global and can be applied wherever an emergency occurs, and notably in areas where terrestrial infrastructure is limited. Such systems are already used in this way, providing imagery and value-added maps (via the International Charter for Space and Major Disasters) as well as vital communications links.

There is scope for improvement, in particular for monitoring, in a continuous manner, areas where natural disasters occur frequently and for providing up-to-date information to the right local authorities. However, space-related agencies do not have the resources to sustain the overall architecture financially (including payment for value-added products needed by third parties, *e.g.* 3–D maps, from value-adding firms), and commercial data providers cannot be expected to provide their products and services at no charge on a continuous basis.

How

- By strengthening international co-ordination efforts already under way, building on existing international programmes that already provide operational assistance (*e.g.* the International Charter for Space and Major Disasters).
- By putting the architecture on a sustainable financial footing. This could be done via the setting up of dedicated funds for disaster management.
- By encouraging greater co-ordination of the different public and commercial emergency systems and extending the scope of their activities to the implementation of prevention measures.

Examples

- *The International Charter mechanisms:* Currently, through the International Charter for Space and Major Disasters, an authorised user can call a single number to request the mobilisation of the members' (*e.g.* space agencies, NOAA) space and associated ground resources to obtain data and information on a disaster at no charge or for a very low fee (usually the costs of reproduction). Each member bears the costs of providing the data and value-added products (*e.g.* maps) to the final users. As the Charter's efficiency improves (reduced time to obtain data and maps) and its services are increasingly used by third parties (such as UN agencies), the organisations in charge of the systems may not be able to cope with and fund the increased activity, especially if there are pressures to make their observation systems work for prevention as well as for emergencies (see Chapter 2).

- *Combining technologies for disaster management:* In times of disaster, a combination of mobile communications, location tools (*e.g.* GPS devices) and Earth observation is essential for co-ordinating emergency teams in remote or hard-to-reach areas. Today's emergency technicians already use some of these applications, often individually (*e.g.* satellite phones). The recent REMSAT Project (Real-time Emergency Management via Satellite), initiated by the European Space Agency, has shown for instance the advantages in combining space technologies as well as the current gaps (see Chapter 3).

- *Towards dedicated funding for disaster management?* The United Nations Action Team on Disasters, which started work in October 2001 following UNISPACE III, recommended in late 2003 the establishment of an international space co-ordination body for disaster management, provisionally identified as the Disaster Management International Space Co-ordination Organisation (DMISCO), with dedicated international funding to ensure sustainable resources for supporting international disaster management efforts, particularly in the developing world. The aim would be to have a fully functional DMISCO within three to five years if the United Nations General Assembly agrees (see Chapter 4).

Recommendation 4.2: Encourage the use of space applications for monitoring international treaties

General view

As the volume of trade, investment and communication across borders rises, and as individuals become more mobile, nations are becoming increasingly interdependent. At the same time, problems of a global nature – such as those related to the state of the environment – are gaining in prominence. A consequence of these developments is that the scope for independent action by governments is being reduced and the effective solution of such problems increasingly calls for co-ordinated efforts at international level. There is therefore a growing need to extend the ambit of international law, so as to promote a more stable and supportive regime for international relations that encourages co-operation and reduces international tensions.

Such progress in international law often takes the form of treaties under which nations agree to abide by certain rules. However, to be successful, the application of such treaties needs to be properly monitored and enforced.

Why

Many international treaties are not fully complied with because their implementation has not been effectively monitored and enforced.

Space-based systems can contribute data for international treaty compliance and verification (*e.g.* environment treaties, negotiation of peace agreements, arms control and disarmament treaties), although external examination of a country's resources or activities from space may raise some delicate political questions relating to national sovereignty.

Current observation systems already provide a unique capability for global observation (*e.g.* optical and radar systems) which is indispensable for effective monitoring of a number of international treaties. This capability will increase in the future, as EO technology (*e.g.* number and diversity of sensors) improves further over the next decades, while the cost of EO systems declines and their use becomes widespread.

How

- By encouraging space agencies to strengthen their partnerships with the secretariats of international treaties and conventions, notably those relating to the Earth's environment and sustainable development, to ascertain how space-based solutions might best be used for treaty monitoring and enforcement.
- By encouraging space agencies to ensure that future satellite missions take fully into account data needs related to monitoring of treaties.

Examples

- *Monitoring the Common Agricultural Policy (CAP)*: In the European Union, satellite imagery and navigation systems will increasingly be useful in order to monitor the application of the Common Agricultural Policy (see Chapter 2).

- *Exploring the use of space for treaty monitoring – Treaty Enforcement Services using Earth Observation (TESEO)*: As a first step in exploring how current and future Earth observation systems may help in the implementation of international environmental treaties, ESA set up in 2001 the TESEO initiative and worked with various treaty secretariats to develop satellite-based services that met their operational requirements. International treaties addressed by TESEO include the 1971 Ramsar Convention on Global Wetlands, the 1992 Kyoto Protocol to the United Nations Framework Convention on Climate Change, and the 1996 United Nations Convention to Combat Desertification. The initiative also aims to increase general awareness of how satellites can help in environmental monitoring. It might build on the successes in using space-based systems to assess the ozone layer, especially the Antarctic ozone hole. The use of space assets supports notably the 1985 Vienna Convention for the Protection of the Ozone Layer, and its Montreal Protocol with its subsequent amendments (see Chapter 2).

- *Monitoring the state of our cultural heritage – the 1972 World Heritage Convention:* At the 16th session of the UNESCO General Conference in November 1972, a number of countries committed to the conservation of world sites of outstanding universal value, from the point of view of culture, history, science, conservation or natural beauty, joined forces to adopt the World Heritage Convention. Today, the Convention, administered by UNESCO, is a success, with 164 Parties to the Convention. As space can be used to monitor the application of the Convention, UNESCO and ESA have agreed to undertake a joint initiative to demonstrate the application of EO and other space technologies (*e.g.* navigation and positioning, communication) in support of the goals of the World Heritage Convention, and to establish a framework of co-operation, open to space agencies and other organisations (see Chapter 2).

> **Recommendation 4.3: "*Special Focus*" Encourage the use of space applications to foster social and economic development in low-income countries**

General view

Many developing nations could use space applications in support of their national economic development programmes. Space applications can be powerful tools for improving the quality of life of citizens in low-income countries and for contributing to the fight against poverty. But international co-operation is necessary to provide more equitable access to space technology.

Why

Space applications facilitate access to information and the management of natural resources. In particular, satellite telecommunications provide the backbone for telehealth and distance learning programmes in areas where no or limited communications infrastructure exists.

Many of the countries that might benefit the most from space applications may not have the means to invest heavily in indigenous or imported space and ground systems. In such cases, development assistance is the best way to ensure such benefits.

Developed countries already agree to provide some level of economic, financial and humanitarian assistance to low-income countries through various bilateral and multilateral mechanisms. It is in the best interests of both providers and recipients that such assistance is delivered as effectively as possible, including by using space-based solutions when appropriate.

How

- By contributing to the training of new users of space applications in developing countries.
- By facilitating the use of existing systems in the developing world, *e.g.* by providing grants for the leasing of transponders from commercial operators and the acquisition of the necessary complementary ground equipment.
- By encouraging international organisations such as UNESCO, the World Health Organisation and the Food and Agriculture Organisation, to promote the use of space for the provision of distance education, telehealth services, the management of natural resources and agriculture, when it is cost-effective to do so.
- By sharing with developing countries the experience acquired in the use of space solutions for the delivery of public services.
- By helping developing countries to participate fully in international efforts to establish a global Earth observation system, both regarding the use they can make of the system and the contribution they can make to the data collection effort.

Examples

- *Building on current programmes:* As the third phase of the project demonstrated, many development programmes using satellites as communications means are under way in several countries, promoted by various organisations, including space agencies (*e.g.* testing by CNES of a dedicated terminal for telemedicine – *station portable de télémédecine par satellite* [SPTS] – in tropical areas) and non-governmental organisations (e.g. the Satellife network provides access via satellite to medical libraries and is used to exchange medical-related e-mail, notably in Africa) (see Chapter 2).

- *The DMC initiative:* The Disaster Monitoring Constellation (DMC) brings together space-related organisations from five countries that are regularly affected by major disasters (e.g. earthquakes, floods): Algeria, China, Nigeria, Turkey and the United Kingdom. Using microsatellites, the DMC constellation is an innovative and cost-efficient way for participants in developing countries to create independent space capabilities (satellite and ground station). DMC should also help to improve the global, systematic and accurate coverage of the planet in co-operation with other systems (i.e. International Charter for Space and Major Disasters) (see Chapter 2).

- *Using space for delivering health services in developing countries – the Health Channel:* The Health Channel will be a satellite broadcast channel to deliver free education to patients and health-care workers in clinics and hospitals in South Africa. It was created through a public-private partnership between the South African Department of Health, Sentech, a provider of broadband communications services, and Mindset Network, a partnership led by Liberty and Standard Bank Foundations (see Chapter 2).

> **Special Focus:** Developing countries might benefit the most from space applications. However, they do not have the means and the expertise to do so. Allocating a fraction of development assistance to promote the use of space systems in the developing world could be a cost-effective way to foster economic and social development.

BLOCK III

Encourage Private-sector Participation

While space activities were essentially public at the beginning of the space age, the role of private actors has expanded in recent decades. First, private actors have been able to exploit successfully, in some markets, technologies that were originally developed in co-operation with or for the public sector. This is notably the case for telecommunication satellites. Moreover, the end of the cold war has created an environment more conducive to the commercial exploitation of space. In a more open world, space firms have been able to restructure and form new alliances, while the opening of markets has benefited important segments of the industry. These commercial developments have also led in many cases to the development of more cost-effective solutions for addressing important societal issues using space technologies (*e.g.* telecommunications networks in remote areas, Earth observation high-resolution data for disaster management).

In spite of such progress, the development of commercial space remains fragile. First, costs continue to be high in the upstream segment of the industry (i.e. space asset manufacturing and launching services) and it is still highly dependent on governments. Second, the development of the downstream segment (i.e. space applications such as satellite communications services, Earth observation services, satellite-based navigation services) is unequal. Some components remain underdeveloped (*e.g.* Earth observation) despite years of efforts, while others (*e.g.* navigation) – although promising – are still at an early stage of development or under public control.

In order to overcome some of these weaknesses, governments need to take action to ensure that private actors are in the best possible position for developing new innovative applications that contribute fully to the economy and society at large. Moreover, governments should fully take advantage of the expertise and resources of private space actors for the development and operation of space infrastructure.

In this regard, three complementary sets of recommendations are presented below:

- **Pillar 5:** Create a supportive legal and regulatory environment for commercial activities.
- **Pillar 6:** Strengthen private provision of space goods and services.
- **Pillar 7:** Foster a more supportive international business and finance environment.

Pillar 5: Create a supportive legal and regulatory environment for commercial activities

General view

The effective operation of our modern economies requires the existence of a legal and regulatory framework that is both stable and predictable. It should provide for clear rules of the game, enforced in a consistent, fair and transparent manner. The framework should also help ensure that entrepreneurship and innovation are rewarded, that barriers to entry and the burden of regulations are minimised, that rent-seeking behaviour is discouraged and that property rights are protected nationally and internationally

Why

These general conditions are not always met in the space sector. First, many countries do not have space laws or only have embryos of such laws. This can be a source of uncertainty for business, as states have ultimate responsibility for interpreting international law and for defining, on that basis, the legal environment that govern the activities of national firms.

Second, many of the space laws that have been enacted are not business-friendly. They often impose stringent requirements and sometimes discriminatory treatment on non-national space companies (*e.g.* disparate fiscal obligations, restrictive operating conditions). Moreover, inequitable licensing regimes and constraining export regulations are major impediments to the commercial operations and competitiveness of many firms in the space sector.

Third, as was apparent in the third phase of the project, the development of space applications also depends on legal and regulatory provisions far beyond the scope of space law *per se*, notably laws that relate to liability, intellectual property, competition and international trade.

Finally, the international regime that governs space is a public law regime that is not fully suitable for private activities. Although this has not yet represented a major impediment to the development of commercial space, it may become one as commercial space activities expand.

6. MAIN FINDINGS AND RECOMMENDATIONS

How

- *Develop national space laws:* This should be done as much as possible in a co-ordinated manner across countries so as to ensure that national legislation is fully compatible across international borders (Recommendation 5.1).
- *Make existing space laws and regulations more business-friendly:* In many cases, this will require a careful balancing of economic and social policy objectives against other public policy objectives, notably security and national defence (Recommendation 5.2).
- Review the impact of the application of general laws on the development of space activities: This needs to be done on a case-by-case basis (Recommendation 5.3).
- *Adapt international law to business needs:* This is a long-term effort that should balance the need for flexibility against the need for more certainty regarding the definition of some basic concepts (*e.g.* defining the boundary between air space and outer space) (Recommendation 5.4).

> **Recommendation 5.1: Develop national space laws if they do not exist or complement existing laws**

General view

Given the liability implications under the international law regime of space activities, it is in the best interests of space-faring nations to implement national space laws in order to regulate the space activities that fall under their jurisdiction. Moreover, national space laws represent a major element of the legal and regulatory environment in which private space actors operate. They establish clearly for them how their national government interprets the international law regime, making the rules of the game more transparent. As legal and regulatory uncertainties are reduced in this way, such actors are in a better position to make sound business decisions.

Why

Without national space laws, private space actors may be reluctant to invest in space ventures, unless the state offers them particular guarantees and incentives. However, such *ad hoc* arrangements are likely to be considered arbitrary and discriminatory, and are not sustainable over time. Indeed, they are bound to create a situation in which the rules of the game are opaque and uncertainties are high, discouraging further private-sector participation in the sector. Moreover, firms with special arrangements with the state may find it hard to forge international alliances with foreign firms operating on a more commercial basis. They may also be subjected to retaliatory action in international markets if foreign competitors object to the special arrangements.

How

- By implementing national laws that cover a number of items of particular importance to the business community: the authorisation and supervision of space activities, the registration of space objects, indemnification regulations, additional regulations (*e.g.* regulations related to insurance and liability, the environment, financing, patent law and other intellectual property rights, export controls, transport law, dispute settlement) as well as procedures for implementing the regulations.
- By better co-ordinating national space laws across countries, so as to facilitate the operations of private space actors at international level. This could be achieved by using a model law for guidance in the formulation of national laws. Such a model law is currently under development by international legal experts and has already received wide acceptance at the international level.
- By taking into account the longer-term effort to adapt international law to business needs (see Recommendation 5.3).

Examples

- *Enacting a space law to attract business:* Although it is not well known, Australia has had quite a long history of involvement in space projects and space launches, mostly military, starting in the 1960s and involving American and British agencies. While it was the fourth nation to launch a satellite in 1967, Australia only drafted and passed its Space Activities Act in 1998. This national space law regulates commercial launches of satellites from Australian soil (*e.g.* liability issues, licensing requirements), opening the door to potential entrepreneurs and larger commercial space firms interested in the unique geographic characteristics of Australian spaceports (see Chapter 5).
- *Space laws currently under development:* Several states are developing national space laws at this time, including France, Germany, Belgium, the Netherlands and Korea (see Chapter 5).

6. MAIN FINDINGS AND RECOMMENDATIONS

> **Recommendation 5.2: Make existing domestic space laws and regulations more business-friendly**

General view

Many of the domestic laws and regulations that affect private space actors are not very business-friendly. A main reason is that many were implemented between 1967 and 1972, with a view to government security interests rather addressing economic or social problems.

The geopolitical scene has evolved considerably over the last few decades, as former enemies have become allies. First, the end of the cold war has reduced international tensions between East and West, alleviating some security concerns and fostering greater co-operation between former enemies (*e.g.* space co-operation between the United States, Europe and Russia). However, new threats have emerged, linked notably to the rise of global terrorism, while dependence of the military on space assets has increased, creating the need to protect these assets.

Why

While legitimate security concerns should clearly override commercial considerations, it is nevertheless important to subject existing measures that have adverse economic consequences to particular scrutiny. As the international situation evolves, and as private space activities expand, existing national laws and regulations that govern space may become less relevant, and security-related restrictions in particular may sometimes no longer be necessary. Moreover, when such restrictions (*e.g.* export controls) encourage other nations to develop their own technologies so as to reduce their dependence on the restricted items, restrictions may do more economic and security harm than good. Consequently, they should be applied with great care and only when the longer-term strategic advantages are clear-cut.

How

- By reviewing national space laws regularly in terms of their impact on the business community, with a view to ascertain whether the constraints on business that were originally put in place are still necessary, or whether other means may be used to reach the same policy objective.[1]

Example

- *The economic cost of the regulation of the international transfer of sensitive technologies:* Several US commercial firms (*e.g.* satellite manufacturers, component suppliers and other exporters of technology products) have complained about lengthy administrative procedures related to the application of ITAR (International Traffic in Arms Regulations) since 1999. In their view, delays in export approval or prohibition of exports put them at a disadvantage compared with their European or Asian competitors. Moreover, the ITAR-related restrictions have induced space systems manufacturers in Europe and Asia to develop new technologies in order to reduce their dependence on US-made components by creating ITAR-free products. Hence, the application of ITAR has fostered import substitution activities which are wasteful from an overall economic perspective and defeat the purpose of the export regime (see Chapter 5).

Recommendation 5.3: Adapt international space laws to business needs

General view

International space law was not drafted with the business community in mind. It is a public law regime that formulates obligations with which sovereign states have agreed to comply in the conduct of their space-related activities. To the extent that there is no formal international body of law that establishes how this public law regime applies to business activities, private firms that engage in space activities face a certain degree of uncertainty.

The uncertainty is somewhat reduced, as noted above, when national space laws are implemented, since such laws establish how states interpret their international obligations and how such obligations apply to their nationals, including national space firms. Even then, however, some degree of uncertainty remains, notably for firms that operate at international level, since their own national law is only one of the elements that would be taken into account in case of an international dispute and reference to the international regime would inevitably be made.

Hence, although the international regime has proved quite flexible over the years in accommodating the development of new applications, some reform of the existing system would seem desirable in the coming years.

Why

Governments clearly have a collective responsibility for developing the legal regime relative to international space activities. As the scope of commercial activities increases over the coming decades, it will become increasingly important to provide a regime that is stable, predictable and fully take into account the need of private space actors.

How

- By using protocols for adapting the existing legal regime to business needs. This would have the advantage of leaving the current space treaties untouched, a major consideration given the traditional reluctance of states to move towards creating new binding international instruments.
- By adopting separate instruments where necessary so as to give a more precise meaning to certain aspects of the treaties or treat specific items. This could take the form of principles and guidelines, codes of conduct or United Nations General Assembly resolutions.

Examples

- The lack of clarity of some terms opens up room for interpretation and potential conflicts for the increasing number of commercial firms involved in space activities. As an example, the Outer Space Treaty does not define precisely "outer space", and no formally accepted legal delimitation of outer space exists at present. Where does space begin? This may have implications for the licensing of future launchers, in terms of whether they are governed by air or space law, and the liability of private entrepreneurs when travelling through different airspaces (see Chapter 5).
- Over the last few years extensive discussions have taken place in UNCOPUOS (United Nations Committee on the Peaceful Use of Outer Space) regarding the possibility of developing a tighter and more workable definition of the term "launching state", as it is crucial for international liability issues (*e.g.* issues raised by the creation of Sea Launch) (see Chapter 5).

Recommendation 5.4: "*Special Focus*" Review the application of general laws and their impact on the development of space applications

General view

In most space applications, the space segment represents only a small, albeit essential, component of the value chain. This means that the laws and regulations that affect other segments of the chain and the final products or services will have at least as much, if not more, of an incidence on the economic feasibility of a particular space application than space law *per se*.

This was clearly illustrated in the third phase of the project, which focused on the development of business models for particular applications. It was noted that the way in which some generic legal concepts are applied in specific cases plays a particularly important role for the success or failure of the applications.

For instance, uncertainties related to liability may have a major bearing on applications as diverse as telehealth services, location-based services or space tourism. As well, it was found that intellectual property issues play a major role, not only for the applications noted above, but also for satellite-based entertainment and Earth observation. Another interesting result is that issues related to the application of competition law are particularly important for Earth observation, satellite-based entertainment and location-based services.

Why

Given the overall objectives of public policy as they relate to space activities, it is the responsibility of governments to review laws that bear on the development of space applications and the operation of space systems in order to ascertain whether these laws are fully supportive of public policy, whether they need to be amended, or whether the way they are applied to space-enabled activities should be modified.

How

- By reviewing the impact of general legal provisions (notably those related to liability, intellectual property, the application of competition law and equitable access) on each major space application, so as to ensure that they do not create artificial barriers to entry or unduly discourage the relevant activity.

Examples

- *Liability and telehealth:* In the nascent field of telehealth, consideration needs to be given to protecting health-care and telecommunications entities from being subject to undue liability. Health personnel who make use of telehealth facilities to provide services to their patients may be sued on the grounds that they did not follow "established practices". This additional risk for health professionals tends to discourage the deployment of telehealth services, even in situations where such services clearly bring significant benefits to the community (see Chapter 5).
- *Intellectual property and Earth observation:* In the field of Earth observation, there is uncertainty in some quarters regarding whether satellite images can be copyright-protected. Moreover, with the digitalisation of data, it is increasingly difficult to protect the intellectual property rights and/or proprietary rights of Earth observation data producers, and to control how the data are distributed (*i.e.* data integrity) or used (scientifically, commercially). Raw data produced by satellites' sensors are normally not protected under intellectual property law, and collections of such data, which have not been subjected to selection or arrangement, are not deemed to be original and are only protected in some countries. Regarding access to data, agencies and private operators alike have their data policies (see Chapter 5).

6. MAIN FINDINGS AND RECOMMENDATIONS

> **Special Focus:** Throughout the project, especially during the third phase which dealt with business and economic models, it was found that national legal and regulatory frameworks – not specific to the space sector – had an impact over a broad spectrum of space applications. This could be a serious impediment to the future development of space applications, hence of the space sector in general.

Pillar 6: Strengthen private provision of space goods and services

General view

Empirical work conducted at the OECD and elsewhere suggests that countries that largely leave the production of goods and services to the private sector tend to perform better, on balance, than those that do not. For instance, a comprehensive review of the literature presented in the 2003 OECD report on *Privatising State-owned Enterprises* concluded that, on balance, there is overwhelming empirical support for the notion that privatisation brings about a significant increase in the profitability, real output and efficiency of privatised companies. An additional interesting finding, presented in another 2003 OECD report on the sources of economic growth in OECD countries, is that the channelling of R&D resources directly to the business sector has a positive effect on innovation, as private-sector firms may be better able to allocate resources towards R&D activities with high commercial return.

Leaving production to the private sector allows the public sector to concentrate on what it does best, *i.e.* providing public goods and services to the population at large and elaborating and enforcing effective rules of the game for private actors.

As in other segments of the economy, the private sector has assumed a growing role in space-related production activities in three major ways: i) through the contracting out by public agencies to private actors of support functions previously supplied internally; ii) through the privatisation of public bodies in charge of developing space assets and operating particular space applications (*e.g.* Intelsat, Immarsat); and iii) through the creation of public/private partnerships (*e.g.* Galileo).

While this process has been reasonably successful to date, it has not been as extensive as in other sectors of the economy. One important consideration in this regard is the dual civil/military use nature of space technology and the fact that governments want to maintain control over the production of technologies they consider strategic.

Why

Further extending the role of the private sector in the production of space goods and services, when appropriate, would bring net benefits to society at large. It could also encourage the entry of new space actors and foster innovation.

How

- To strengthen contracting out to the private sector (Recommendation 6.1).
- To move partially or totally to the private sector activities of a commercial nature, if economically viable (Recommendation 6.2).
- To encourage entrepreneurship and innovation (Recommendation 6.3).

Recommendation 6.1: Foster public procurement from the private sector

General view

As noted in the first phase of this project, public expenditures on space represent a major market for the space industry. About 70% of such expenditures are indeed purchases in one form or another from the industry. They include products and services for R&D purposes, space hardware (including orbital infrastructure) and the procurement and operation of launchers. For instance, in 2003, government-funded missions accounted for 75% of the 63 launches performed worldwide.

Two types of public customers can be identified: space agencies, which focus mainly on R&D and therefore develop new products (i.e. product definition is open) and institutional clients that typically acquire products off the shelf.

Why

In appropriate circumstances, contracting out may offer public agencies a number of advantages. First, it may free up resources to focus on how the service or product may best be applied, rather than on the day-to-day production of the service. Second, it may provide access to the contractor's knowledge, network and research. Moreover, in a competitive environment, it allows the agency to choose the product or service that best suits its needs.

Contracting out may also be beneficial for the contractor. First, it helps to increase its revenue base. Public procurement may also allow the firm to reap economies of scale and of specialisation. Finally, it may allow the firm to diversify its sources of revenue, a particularly significant advantage when private demand tends to fluctuate significantly, as is the case for space.

6. MAIN FINDINGS AND RECOMMENDATIONS

Public procurement can also be a way to stimulate the entry of new innovative private players into the industry, notably through procurement programmes specifically targeted to small and medium-sized enterprises (SMEs).

However, it is important to keep in mind that contracting out is not appropriate in all situations. It needs to be considered on a case-by-case basis, notably when the supply industry is highly concentrated.

How

- By establishing clear guidelines for public provision. In this regard, particular attention should be given to the costs and risks involved, notably: i) the complexity of the contractual arrangements that need to be established; ii) the significant barriers to exit, once a long-term relationship has been forged with the contractor; and iii) the high financial risks that may be involved. These problems are particularly important when only a few firms are in a position to meet contractual requirements and the service or product required is very specialised.
- By putting in place mechanisms for encouraging the participation of innovative SMEs in the procurement process (*e.g.* reduce the paperwork required for smaller contracts; request prime contractors to allocate a share of their contract to SMEs or make it a proposal evaluation criterion; setting aside a share of the procurement budget for SMEs).

Examples

- *Establishing clear guidelines for public provision of space products and services:* The European Space Agency, as an international organisation with 15 member states, has a detailed framework for the provision of commercial good and services. The procurement rules are based on three main instruments: the ESA's Council Contracts Regulations, the Industrial Policy Committee Terms of reference and the General Clauses and Conditions for ESA Contracts. These regulations allow private actors, selected after competition in most cases, to know well in advance their rights and obligations as contractors (see Chapter 3).

> **Recommendation 6.2: Privatise commercially viable business-oriented government activities**

General view

There is overwhelming evidence that privatisation has had largely positive effects on incentives, profitability and performance of privatised enterprises in general. It is clear, however, that privatisation should not be proposed across the board, as a strong case can be made for a number of activities to remain mainly public (*e.g.* weather satellites). In that context, "business-oriented activities" are defined here as activities designed to offer goods and services for sale to essentially private customers on a commercially viable basis (*e.g.* the provision of telecommunications services to business users and the general public). As discussed in the context of Recommendation 6.1, the case for privatisation is not as strong when the "customer" is only the government, since in that case, the government may still have to assume most of the risks involved.

Why

Space activities have been carried out in the public sector for a long time. As commercial space expands, it is only natural that some of these activities become candidates for privatisation. This has a number of advantages. First, it brings a business discipline to production activities, as private firms have a strong incentive to keep costs down and to produce goods and services buyers are willing to buy. Moreover, it allows private capital in production, capital that would not otherwise be attracted. Finally, it gives the privatised firm more flexibility to pursue international markets and seek international partnerships that enable it to focus its activities in areas in which it has a comparative advantage.

How

- To the extent possible, business-oriented activities in governmental agencies that serve private markets should be privatised.
- Monopolies may also be privatised, but a regulatory system and public service obligations on privatised companies may be needed to serve the public interest.
- Consideration should also increasingly be given to public-private partnership (PPP) schemes when appropriate, especially for long-term projects in which the infrastructure to be developed jointly can serve both public and private needs.

Examples

- *Inmarsat, a successful privatisation:* In the 1990s, several intergovernmental satellite operators, with increasing commercial activities, were privatised successfully (Intelsat, Inmarsat). Created in 1979, Inmarsat was a maritime-focused intergovernmental organisation, whose purpose was to improve maritime communications and radiodetermination capabilities of ships at sea (in particular assuring distress and safety of life at sea communications).

Owing to its increasing commercial activities, governments decided to privatise Inmarsat and created two complementary bodies in April 1999: a limited company, which has since served a broad range of communications markets, and the International Mobile Satellite Organisation (IMSO), an intergovernmental body established to ensure that the Inmarsat company continues to meet its public service obligations, in particular its Global Maritime Distress and Safety System (GMDSS) obligations (see Chapter 4).

- *Public-private partnerships (PPP) in the space sector:* There are several options for developing PPP in the space sector. For instance, the UK Private Finance Initiative (PFI) model used over the past two decades in several key infrastructure programmes, including for hospitals, prisons and roads, was recently adapted to the space sector. In 2003, the PFI approach was used for the first time to provide military satellite communications services (Skynet 5 programme). The British Ministry of Defence (MOD) selected (after competition) Paradigm Secure Communications as the operator for a 15-year contract period to build and operate the military communications systems. This PPP provides the MOD with significant benefits: no large investment up front, risk sharing with the private operator, use of system capacity as and when required. The operator benefits include in particular guaranteed revenues for a long period (they could be worth more than GBP 2.5 billion to 2018), and possible additional profits by reselling available capacity to MOD-approved defence and other governmental users from overseas countries and multinational organisations. Already, several NATO countries are planning to sign up for Paradigm services and the Paradigm model could be extended to other regions of the world for military forces that prefer to purchase capacity rather than own their own satellites, and cannot rely solely on standard commercial spacecraft (see Chapter 4).

> **OECD follow-up:** The increasing impact of public private partnerships in the commercial space sector has not yet been fully analysed. PPP might represent a useful policy tool to increase private involvement and investment in the development of space systems. Diverse approaches have been adopted in OECD countries for major infrastructure projects. They should be thoroughly analysed in order to draw pertinent lessons and identify best practices for future partnerships in the space sector.

> **Recommendation 6.3: "Special Focus" Encourage entrepreneurship and innovation**

General view

Economic analysis strongly supports the view that innovation is a major contributor to economic growth. For instance, the 2003 OECD report, *The Sources of Economic Growth in OECD Countries*, points out that, when comparing economic performance across OECD countries, the development and diffusion of innovation and new technologies make an important difference for growth prospects. The report also finds that competition and innovation are closely linked and that, indeed, pro-competitive regulations help growth by promoting innovation. A related finding is that the entry of new firms in a sector tends to boost productivity. In this overall context, SMEs play a major role. They constitute an important and dynamic element in all economies as they drive innovation, especially in knowledge-based industries.

Why

Governments have a responsibility to encourage the development of new products and services that generate revenue and jobs. Although they should not pick winners, they should be supportive to innovation.

The space industry is highly concentrated and there are strong barriers to entry. This helps to stifle competition and has a detrimental impact on innovation.

In comparison with other industries, the number of truly innovative and independent SMEs in the space sector remains relatively small. Space-related research and technological development are quite expensive, and SMEs have difficulty in accessing appropriate funding schemes or in benefiting from appropriate technology transfer with which to create new products and services.

How

- By setting up nationally a business environment that is conducive to innovation and entrepreneurship (*e.g.* a tax system that entails low compliance costs; the transparent and equitable application of rules and legislation; simple and transparent licence and permit systems; efficient bankruptcy laws and procedures; understandable and coherent product standards in world markets; clearly defined property rights; fair and reasonably priced dispute settlement procedures; and light, predictable administrative procedures).

- By encouraging space entrepreneurs, with no vested interest in the existing space infrastructure management to develop and try new technological ideas and innovative space operations. These efforts might include the setting up of special prizes, reserved to SMEs and funded by agencies and larger private organisations, based on the X-Prize model.
- By supporting space entrepreneurs who are attempting to develop new innovative applications (*e.g.* space tourism).
- By encouraging SMEs to participate in large space programmes as contractors.

Examples

- *Lessons learned from the X-Prize:* The X-Prize is a private initiative that stimulated substantial efforts by private entrepreneurs to develop a sub-orbital plane with no or only marginal support from the government. The prize was won in September 2004 by SpaceShipOne, the first private suborbital plane to reach space. Its success and the efforts of other space entrepreneurs could pave the way to the development of a job and revenue-generating space adventure/tourism industry in the coming years. These space entrepreneurs may also contribute to the development of new flight technologies and cost-efficient operations mechanisms (see Chapter 3).
- *Programmes involving SMEs:* Several space agencies' programmes are already reserved for SMEs or include specific requirements for large contractors to partner with SMEs. Dedicated initiatives such as those of the European Space Agency support SMEs that possess new technologies with potential applications in the space sector and promote the development and diversification of SMEs already working in that sector (see Chapter 4).
- *Encouraging the creation of a new private satellite operator – SES Global:* In most cases satellite operators were originally public and have been gradually privatised. However, some were private from the start. SES Global is an interesting case in point. The company, entered the space arena in 1985 with an initial small investment and a state guarantee. By 2004, the company had become the world's biggest satellite operator, with a fleet of 40 satellites across the globe (see Chapter 4).

Pillar 7: Foster a more supportive international business and finance environment

General view

An open business environment at international level contributes to overall economic development by fostering a more efficient allocation of resources, by encouraging the introduction of new innovative products and services which can take advantage of a larger market base, and by facilitating the rapid diffusion of new technologies.

Because the production of space-based services is typically characterised by increasing returns to scale, it is vulnerable to regulations that tend to fragment markets.

Why

As illustrated by the success of the World Trade Organisation's (WTO) Basic Telecommunications Agreement on trade in telecommunications services in 1997, the provision of space-based services can benefit significantly from a more open international business environment, given the ubiquity of such services.

In spite of these advances, many markets for space-based products and services remain fragmented along national lines and there are severe constraints on movements of capital. Hence, there are still significant potential efficiency gains to be achieved through further liberalisation that only governments can undertake.

How

- By extending the liberalisation of trade to a broader range of space-based services (Recommendation 7.1).
- By encouraging the establishment of international standards (Recommendation 7.2).
- By improving the allocation of spectrum and orbital positions (Recommendation 7.3).
- By encouraging the private financing of space activities (Recommendation 7.4).

> ### Recommendation 7.1: Extend free trade discipline to open up markets

General view

The provision of space-based services is best suited for large markets, given the broad geographical coverage that satellites offer and the very low marginal cost of providing such services to an extra customer. The production of space hardware is also subject to significant economies of scale. This is because fixed costs relating to R&D are a very important component of total costs, given the complexity of space technology and the short production runs that typically prevail in the industry. Hence, such production, too, would benefit from access to a large market.

However, market access is restricted. For strategic/security reasons, public procurement – the largest segment of the market for space goods and services – is often limited to national or regional level. This tends to stifle competition and result in a misallocation of resources, as firms that should exit the industry remain artificially active, while firms that could serve the market more efficiently are barred from doing so.

Why

Governments are mainly responsible for the imposition of these restrictive rules. At international level, the relevant rules of trade agreements, as monitored by the WTO, should be extended as much as possible to all space-related activities of a commercial nature. Lack of clear rules of the game can result in conflicts, notably regarding international trade in satellites and launch services. Moreover, as the importance of commercial activities increases, the cost imposed by trade restrictions (*e.g.* export controls) is likely to increase and become counter-productive from a strategic perspective.

How

- By reducing the regulatory burden on non-governmental launch activities through the conclusion of mutual recognition agreements. Such agreements, while giving effect to the legal obligations and interests of affected launching states, would allow the acceptance of the authorisation granted by other launching states.
- By promoting the development of a single unified (international) intellectual property rights regime applicable to outer space activities.
- By giving the WTO the lead role for dealing with market access issues as well as for dispute settlement related to trade in space goods and services.
- By encouraging, in the case of the telecommunications sector, governments that are members of the WTO to respect commitments undertaken in the Basic Telecommunications Agreement (Protocol 4) and open their markets for telecommunications, in a transparent manner.
- By concluding technology safeguard agreements relating to the launching of foreign satellites, as well as the use of foreign launch vehicles, in order to protect technology that is subject to export control regulation.
- By maintaining an open regime regarding trade in space-based navigation equipment.

Example

- *Liberalising government procurement* – The WTO Agreement on Government Procurement (AGP): As space commercialisation progresses in the future, some efforts undertaken in the context of the WTO might be noteworthy for government procurement of space systems. For instance, the AGP of 1979 has started opening up the business of government procurement to international competition. It is designed to make laws, regulations, procedures and practices regarding government procurement more transparent and to ensure they do not protect domestic products or suppliers or discriminate against foreign products or suppliers. This regime could be gradually extended to space commerce activities (see Chapter 5).

Recommendation 7.2: Encourage international standards

General view

Experience acquired over the years in a large number of economic sectors strongly suggests that the setting up of standards can significantly help to improve productivity and cut costs. Transaction costs are reduced, economies of scale can be achieved, competition is strengthened and systems are fully scalable.

However, in certain circumstances, standards may impede innovation and market development. This is related, at least in part, to the fact that standards have a strategic dimension. Since they are embodied in national or regional regulations, they can be crafted to either facilitate or impede market access.

Standards have extensive repercussions on a company's product design and testing. It has therefore become critical for every competitive company to participate actively in the process that develops space-related standards (*e.g.* equipment in space and ground segments, dedicated applications).

There are currently many competing standards development entities at national and international levels (*e.g.* International Telecommunications Union [ITU], International Organisation for Standardization [ISO], industry).

Why

International standards are not yet fully developed in the space sector for two reasons. First, many standards used in industry in the past were based on military standards and set up independently by space agencies or other technical agencies. Second, the need to standardise countries' space systems is relatively recent. It has resulted from the multiplication of international co-operation projects and from the increasing commercialisation of space-related products and services (communications satellites in particular) at the international level.

It is the responsibility of governments to encourage the development of standards and to ensure that those that are set do not unduly favour particular players.

How

- By encouraging industry to set standards that are open and help to facilitate market access. In this regard, public procurement can provide a vehicle for encouraging industry to tool up and/or develop standard practices.
- By encouraging industry to participate in international standards-setting organisations such as ISO, the Consultative Committee for Space Data Systems (CCSDS) (an international forum for space agencies and space industry) and the ITU.
- By reviewing the competitive implications of existing standards, taking into account the legitimate concerns of companies to protect their intellectual property rights so as not to discourage innovation.

Example

- *Standards and the development of entertainment via satellite:* On the basis of the analysis in the third phase of the space project, standards appear crucial for the development of entertainment via satellite at different levels of the value chain (i.e. establishment of open standards for the digitalisation and delivery of content, as well as for the manufacture of consumer equipment). Significant progress has been achieved towards the development of open standards in recent years for digital broadcasting (*e.g.* development of the MPEG-2, MHP, DVB-S standards). However, uncertainties remain in other areas, notably for two-way communications (*e.g.* the DOCSIS *versus* DVB-RCS debate). Controversy also still surrounds the development of "plug-and-play" standards between direct broadcasting service operators and the cable industry (see Chapter 5).

Recommendation 7.3: Improve the allocation of spectrum and orbital positions

General view

The regulatory regime developed in the context of the International Telecommunication Union (ITU) for the international regulation of telecommunications services has encompassed satellite communications since the early 1960s, in particular regarding technical issues such as frequencies, non-interference and orbital allocations.

As the scope for wireless communications increases, efficient spectrum allocation and orbital allocation will become an increasingly important policy and economic issue. In the context of increased commercialisation, the regulatory process should progressively be improved so as to lead to a more efficient use of the spectrum and orbital slots.

Why

Frequency allocation is particularly important for space applications, since all space-based services depend on the ability to communicate wirelessly. Moreover, many services are provided by geostationary Earth orbit satellites, so that orbital allocation is also important.

Currently, allocations are often made on a first-come-first-served basis although certain *a priori* rules are used for some telecommunications services. Moreover, no ownership is assigned to particular orbital slots, in accordance with United Nations treaties (principles of equal access and non-appropriation of space). This raises a number of issues:

- The interests of developing countries may not be sufficiently protected: because of the first-come-first-served rule, which applies to the allocation of frequency and orbital slot, too few frequencies may be available to meet their future needs.
- The allocation process is inefficient. The current regime encourages paper satellites or filings by entities that do not seriously intend to deploy satellite systems. This increases the ITU's workload and creates uncertainties for actual system operators.
- The ITU lacks authority to make and enforce decisions. This leads to unresolved disputes, permanent use of orbital positions, non-payment of penalties and lack of sanctions for delinquent satellite operators.
- Changing technologies means that the traditional way of allocating frequencies may no longer be appropriate.

How

- By encouraging the participation of private actors. The ITU regime is still essentially a state-oriented public one, even if non-governmental entities have greater opportunities to voice their interests and concerns.
- By giving proper attention to the legitimate concerns of developing countries, while at the same time ensuring the efficient use of scarce spectrum.
- By exploring the feasibility of auctions for the allocation of orbital slots and spectrum.

Example

- *Improving the allocation process:* The ITU has started tackling the paper satellite phenomenon, as it is faced by a large and growing backlog of co-ordination requests, which also slows the development of legitimate commercial systems. The ITU has done so through its regular Plenipotentiary Conferences (the supreme organ of the ITU convened every four years) with new "due diligence" administrative and financial procedures to discourage unwarranted filings. This process obliges all operators and national administrations that are filing requests for co-ordination of satellite systems to provide full details of system contractors, including manufacturers and launch companies, along with a planned schedule of system deployment.

Recommendation 7.4: Encourage the private financing of space activities

General view

In most business activities, the ability to finance the acquisition of productive assets by borrowing from private lenders is essential. Typically, the productive asset is used as collateral so as to protect the lender against default by the borrower.

In the case of the space sector, the range and volume of activities being conducted by private actors have dramatically increased over the last decade. However, commercial space systems are extremely capital-intensive to plan, design, construct, insure, launch and operate, and they can take years to complete. For this reason, work currently being carried out by the International Institute for the Unification of Private Law (UNIDROIT) to provide clear financing schemes for space companies will be very useful for the future of the commercial space sector.[2]

Why

There is as yet no established market for commercial financing of private space activities, as exists for most other industrial sectors. To fill this need, a dedicated space Protocol to the UNIDROIT *Convention on International Interests in Mobile Equipment* (opened for signature in 2001) is being drafted.[3] It would set a framework through which states can support a system of asset-based and receivables financing.

By permitting secured financing for the space sector, the Protocol has considerable potential to enhance the availability of commercial financing for outer space activities and to further the provision of services from space to countries in all regions and at all levels of development.

How

- By supporting UNIDROIT efforts to finalise the space assets Protocol to the UNIDROIT *Convention on International Interests in Mobile Equipment*.
- By signing and then ratifying the future Protocol promptly so as to make it applicable in national law. There is danger that the Protocol will not be effective if many states are reluctant to accept such a regime, or only want to accept it if there are sufficient opt-out possibilities.

Example

- *The challenges of financing a space venture*: Although the main weaknesses of space ventures when they seek financing are important (*e.g.* long lead times for project development, long time to break even and generally high uncertainty or risk), context plays a key role as it does in any industry. In August 1999, the mobile satellite operator Iridium LLC filed for bankruptcy, in part because of a failed time-to-market analysis. This directly affected in turn other ventures and their financing efforts, such as ICO Global Communications which was looking for a minimum of USD 500 million from its existing shareholders and the financial markets, but ICO filed for bankruptcy shortly after Iridium's filing (see Chapter 4).

Notes

1. Questions that can be raised in this context include: Are the measures still needed, given the evolution of technology and international relations? Do they really achieve the expected security objectives? Do they have perverse unintended consequences that may undermine their effectiveness? Are the benefits in terms of enhanced security worth the cost in terms of lost economic activity?
2. UNIDROIT is an independent intergovernmental organisation. Its main purpose is to study needs and methods for modernising, and co-ordinating private and in particular commercial law between states and groups of states to contribute to the development of a reliable and efficient, harmonised international legal framework for public and private actors alike.
3. This UNIDROIT Convention, opened for signature in 2001, already sets universally applicable general rules for the provision of default remedies, the international registration of security interests and the rules of priority regarding such security interests for mobile equipments. So far, 28 states have signed the Convention (October 2004) and its related dedicated Protocol on Matters Specific to Aircraft Equipment (Cape Town, 2001).

ISBN 92-64-00832-2
Space 2030
Tackling Society's Challenges
© OECD 2005

ANNEX A

Case Studies on Selected Space-based Applications

Introduction

This annex provides an overview of the case studies conducted in the third phase of the project in order to explore – in close collaboration with members of the Project Steering Group – possible appropriate economic/business models for the successful development of specific space-based applications.

The purpose was not to identify actual business opportunities, but rather to reach a better understanding of the factors that are likely to play a critical role in the success of the applications under investigation and to map out areas in which government action may be required, so as to provide a more concrete basis for the policy-oriented reflection and the formulation of recommendations in the fourth and fifth phases of the project. Indeed the term "economic/business model" is used throughout to reflect the fact that in a number of cases the application under consideration will be carried out by the public sector to provide a public service rather than to exploit a market potential.

Five case studies were selected: telehealth via satellite, satellite entertainment, risk and disaster management, location-based road traffic management and space tourism. These case studies were chosen by the Steering Group and the OECD Project team because: i) they represent a useful cross-section of public-oriented applications with high social value, private-sector-driven applications with strong commercial dimensions, and applications offering a mixture of the two; ii) they take advantage of space technologies; and iii) they have a reasonably good probability of success in the longer run.

To facilitate comparability and readability, each of the five case studies is divided into five standard sections:

● An introduction gives the rationale used for selecting the application.

- A first section provides a brief description of the main actors and shows how they may interact with one another.
- A second section outlines the major factors that are critical for the success of the application.
- A third considers the economic/business models that can be used for the deployment of satellite-based solutions.
- A fourth explores the prospects for the respective economic/business models, taking into account the three scenarios developed in the second phase of the project and outlined in the Chapter 1 of this volume.
- A final section lists a number of references pertinent to the area concerned.

At the end of the annex, a concluding chapter identifies seven areas in which action is required to support the successful implementation of the economic/business models considered.

1. Telehealth

Introduction

Health spending represents a large and growing share of GDP in most countries. OECD economies devoted on average almost 8.5% of their GDP to health expenditures in 2001, up from around 5.5% in 1970. Given the strong link between rising income levels and rising levels of resources devoted to health care, the upward trend is expected to continue in the coming years, albeit at a somewhat slower rate. There is concern in both OECD and non-OECD countries about the escalating cost of health services and about the limited access to health care experienced by many citizens in more remote regions. These problems are likely to be exacerbated in the coming decades as populations age, not only in OECD countries but also in some of the larger emerging economies.

Telemedicine and telehealth offer the prospect of helping to extend health-care coverage and to reduce health costs. By extending the reach of medical care to populations in remote areas, to disaster assistance teams, ships at sea, etc., they will help overcome geographical barriers to access. Telehealth can offer opportunities for satellite-based solutions in areas where satellites have a comparative advantage over terrestrial technologies. Public and private actors will have an important role to play in these developments.

Telehealth is generally defined as the use of information and communications technologies (ICTs) (including satellite communications) to support long-distance clinical health care, patient and professional health-related education, public health and health administration. It can help to alleviate the problems noted above by providing a broad range of services designed to improving the quality, geographical reach and timeliness of healthcare. Such services include:

- **Long-distance provision of health education and decision support material to professionals and patients.** Health professionals can be kept informed in real time of the most up-to-date protocols and findings. Patients can receive timely information and recommendations made by health professionals.
- **Long-distance administration of health services.** The various health services using the network can be connected and integrated so as to offer a seamless continuum of care to patients.

- **Long-distance diagnosis and treatment.** Telehealth provides a tool for accomplishing several of the tasks involved in disease management. This includes the communication of physician guidelines to the patient, monitoring the health of the patient, patient education and behavioural modification interventions.

The main actors

The provision of telehealth services may involve different configurations of actors. Figure A.1 provides an illustration of what such configurations might look like. In the hypothetical example presented here, the provider of telehealth services is represented in the shaded square. It is assumed to be a joint venture between a satellite operator providing the communications link, affiliated health professionals offering the telehealth services and the telehealth network operator that uses the satellite communications link (and perhaps other ground-based communications links) to transmit telehealth data and that provides support for the operation of the telehealth equipment.

Figure A.1. **Main actors involved in the provision of telehealth services**

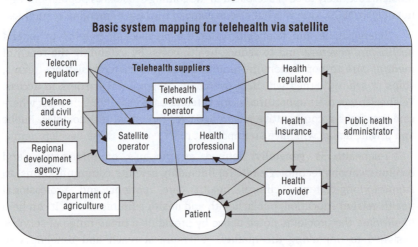

Source: Authors.

Outside the shaded square are represented some of the major actors that may influence the development and operation of the telehealth service. These include first, healthcare institutions which are potentially the main users and providers of telehealth services. Second, payers (public and private health insurance) play a key role since they determine whether telehealth services are reimbursed and how. Third, telecommunications and health regulators govern the day-to-day activities of the telehealth service provider.

Finally, public health administrations may encourage the development of telehealth in order to facilitate access, reduce costs and improve the quality of health care, as well as to foster the dissemination of public health information. Politicians may also encourage the development of telehealth if they can be convinced that it may be able to address broad policy issues, such as inequalities in health care, and that it may reduce costs while delivering at least equivalent, if not better, standards of health care than more traditional alternatives. The funding of telehealth networks can be provided by national authorities, including national telecommunications agencies, regulatory bodies, health agencies and various other national ministries (*e.g.* Agriculture, Defence, Veteran Affairs), as well as regional authorities (*e.g.* states or provinces) and private foundations. In Europe, the funding can also be made through regional development programmes supported by structural funds and through Framework Programmes in the case of research projects.

Critical success factors

Telehealth service providers deal with many different actors and face many challenges. In this context, some factors (technical, economic, legal, regulatory, financial, among others) will be crucial to their success. These can be grouped under two main headings: *i)* the factors that relate to the business model to be used for the development of the application; and *ii)* the factors that shape the environment in which the application is implemented.

Critical success factors for the business model

To be successful, the telehealth application should be so designed as to fulfil the needs of health administrators, health professionals and patients in the most cost-effective manner. Moreover, sources of revenue should be clearly identified and all possible sources of funding should be tapped. More specifically, the application should:

- **Fully meet the needs of health professionals.** It should be user-friendly, save time for clinician professionals and offer diagnostic-quality images, diagnostic-quality sound and a true representation of paper-based information (lab reports, requisitions).
- **Provide effective support at all telehealth sites.** Such support is essential for building confidence in the telehealth network among health professionals and patients alike. This includes equipment, administrative and technical support, clinical support and clinical administrative support.
- **Effectively address security and privacy requirements.** This may require separate secure telecommunications links if encryption is not commercially available (*e.g.* for video telehealth services).

- **Facilitate providers' acceptance.** This may call for approaches to financing that reduce the risk borne by the user (*e.g.* leasing or outsourcing model).
- **Clearly establish how revenues will be generated.** Estimating revenues from a patchwork of public and private reimbursement policies is not easy for the telehealth provider and may require a specialised administrative capacity.
- **Leveraging sources of funding.** The success and sustainability of telehealth networks depend on their ability to leverage capital resources creatively (*e.g.* funds from public sources, private foundations and user fees).
- **Provide for an efficient information flow.** Recognising that poor information flows in health-care settings contribute to gross inefficiencies, inequities and quality variations, business models that focus on telehealth as an enabler of better, faster, cheaper information flows will present better cases for acceptability and profitability.

Critical success factors in the business environment

In addition to factors directly related to the design of the telehealth application discussed above, the telehealth entrepreneur must take into account critical factors in his/her business environment. This includes notably:

- **Licensing.** Because a telehealth network can extend over several jurisdictions, it is important to ensure that health professionals using the network are fully entitled to exercise their profession across all jurisdictions covered by the network or appropriate policies are put in place to meet licensing requirements.
- **Standards.** Lack of broadly accepted standards is a major obstacle to the development of a telehealth network. It restricts interoperability and prevents full exploitation of the capability of the technology.
- **Communications.** High communications cost is often cited as a major obstacle to the development of telehealth network, notably in rural and remote areas.
- **Liability.** Concern about liability suits may induce health professionals to shy away from procedures such as telehealth that are not considered "well established" in the profession. The definition of malpractice is based on the premise that the doctor did not follow widely accepted practices.
- **Technology.** Uncertainties regarding the future development of telehealth hinder private investment in the development of telehealth technologies. Moreover, technology transfer from the military has been slow.
- **Provider acceptance.** The adoption of telehealth is a challenge for health professionals: telehealth applications require technical knowledge and technical problem-solving skills. They also require a new workflow routine

and organisational support. Moreover, if it is not accepted by all (*i.e.* health administrators, health professionals and patients), telehealth is unlikely to work effectively.

Economic/business models

Space-based telehealth services need to take advantage of space's particular strengths. This means that they should focus on three main areas of need: rural and remote populations; populations on the move; populations involved in a disaster.

Model 1: Providing telehealth services in rural and remote areas. The customers here are health professionals and patients. The network is run by an operator, which is set up by a consortium that may include satellite operators, health management organisations and research organisations. The network enables health professionals to provide remotely a number of health services, including health education, diagnosis, treatment and monitoring. The network should meet the critical success factors noted above. The consortium finances the construction and operation of the network. Such financing may involve both public and private funding. Typically, the key players will be the health service providers and the payer, who must accept to pay for telehealth services. Other major players are the public health authorities which have a mandate to promote equal access to health care and may be prepared to encourage financially the development of telemedicine in rural and remote areas.

Model 2: Serving the health needs of people on the move. Given that the clientele for this kind of service is affluent (*e.g.* tourists and executives of multinational enterprises) and emphasises quality of care and convenience, a private business model may be the most appropriate. The main actors might be an enterprising health service provider (*e.g.* a well-known hospital with a strong research and education capability) creating a joint venture with a satellite operator and a telehealth network specialist. The telehealth service would provide the link from the affiliated health service provider and the client, ensure installation and maintenance of the terminal equipment in the client site (*e.g.* cruise ship) and be responsible for training the client's health personnel in charge of using the terminal equipment.

Model 3: Serving the health needs of people in a crisis situation. In a crisis situation when ground facilities are destroyed, including terrestrial telecommunications links, satellite-based solutions may be the only ones able to function effectively. Moreover, satellites can be rapidly deployed over a crisis location. Hence, satellite-based telehealth networks are likely to be favoured by civil protection and national security authorities, at least as a backup to terrestrial ones. One might envisage, for instance, a satellite-based telehealth network linking centres for disaster control to emergency response teams, disaster management assistance teams, and chemical, biological, radiological, nuclear or explosive teams. Such a network would essentially be public and is likely to involve space assets under military control.

Prospects for the future

The limited use of telehealth today is due to technical and institutional obstacles to its implementation, as well as insufficient empirical evidence regarding its cost effectiveness. However, there are reasons to believe that the situation will improve considerably in the coming years.

First, given the growing interest of health professionals, a stronger evidence base is likely to develop gradually, as well as better guidance for overcoming the obstacles to the successful implementation of telehealth systems.

Second, progress in a broad range of converging and complementary technologies should help to reduce significantly the cost of delivering telehealth services and to facilitate the development of a fully fledged telehealth network. This includes not only progress in telehealth technologies *per se*, but also progress in related disciplines, notably health informatics and e-health and more generally, progress in communications technologies (notably broadband).

At the same time as "technology push" may make telehealth increasingly feasible, changes on the demand side will "pull" its implementation. A growing number of individuals are likely to become more mobile, as incomes rise and transport becomes more efficient and there will therefore be a greater need to be able to provide health services wherever that population happens to be. Moreover, as the population in OECD countries ages, there will be an increasing demand for homecare, most of which might be accommodated via telehealth. In addition, growing concern about security will induce health authorities to develop emergency systems that can respond flexibly and effectively in times of crisis.

Fourth, greater concern about providing equal access to health care will drive the development of telehealth networks to serve the needs of individuals living in rural and remote areas or are unable to access health facilities easily. Finally, if the cost effectiveness of telehealth is confirmed convincingly by new empirical research, the greatest driver for telehealth may very well be governments seeking to cut healthcare costs.

On balance, the main conclusion is that there are indeed good reasons to believe that the pace of implementation of telehealth will accelerate in the coming years. Other things being equal, this should create, overall, a favourable environment for the satellite-based applications and the related business models outlined above. When considering the scenarios outlined in Chapter 1, it appears likely that the development of telehealth will be fastest under the Smooth Sailing scenario: the relatively open environment in this scenario is favourable to the implementation of space-based solutions and demand from people on the move is likely to be very high. Moreover, in this rather optimistic vision of the future, telehealth will be a powerful way to extend health care to the developing world and remote areas. In the Back to the Future scenario, the environment is less open and international co-operation is not as extensive, so that from this perspective the prospects for telehealth may not be as good. However, when both civil and military aspects of security are taken into account, greater security concerns may be an important driver for the further development of telehealth. In Stormy Weather, the prospects for telehealth appear to be less promising, notably satellite-based solutions. However, security concerns will remain a major positive driver. Cost considerations might also play a role, if telehealth truly contributes to cut health-care costs. Hence, on balance, the prospects for telehealth are good across all three scenarios, although they appear to be best under the first scenario.

2. Entertainment

Introduction

Satellite entertainment services represent a segment of the huge and dynamic media and entertainment sector. The services are provided by satellite direct-to-home (DTH) platforms and include: i) regular broadcasting services (which represent the bulk of such services today); ii) enhanced (including high-definition television – HDTV) and interactive services (such as interactive television – iTV); and iii) broadband entertainment services provided through broadband access (typically the Internet via satellite).

In recent years, satellite DTH platforms have grown rapidly, with close to 60 million subscribers as of January 2004. Rising subscriptions have been supported by the unchallenged diversity of content offered by satellite. DTH platforms currently broadcast more than 7 200 TV channels, and spend around USD 16 billion in programming a year.

The entertainment sector is currently in a state of flux, as disruptive new technologies, such as the digitalisation of content and the development of broadband access, offer new ways to produce and deliver content and call into question existing modes of operation. For DTH platforms, these developments offer new opportunities but also new challenges – not only to their ability to maintain their strong position in the broadcast distribution segment, but also to their capacity to penetrate other broadband entertainment markets and develop successful business models.

The main actors

The system mapping presented in Figure A.2 gives a general picture of how satellite broadcasting services are delivered and of the main actors involved. The key players are the satellite operators and the DTH platform operators which are either vertically integrated or linked through long-term lease contracts. In some cases, a DTH platform operator may lease an entire satellite for periods of up to 10 or 15 years. Other important actors are the content providers and content aggregators that provide the input to the system, while revenues come from advertising and subscriptions. The government plays a major role in determining the rules of the game, largely through the regulatory agencies.

ANNEX A

Figure A.2. **Main actors involved in the provision of satellite entertainment services**

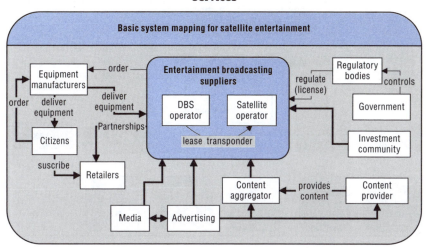

Source: Authors.

In the case of broadband entertainment services (Figure A.3), the configuration of actors is somewhat different: a retailer uses the satellite capacity provided by the satellite operator to offer an end-to-end service to its customers, *i.e.* access services to Internet service providers (ISPs), telecommunications companies or large integrators, which in turn provide services under their brand names to their end users. This retailer may or may not be affiliated with the satellite operator.

Critical success factors

In the selection of applications and in the designing of the business model to be used for their development, entrepreneurs must take into account a number of factors that will be essential to their success.

Critical success factors for the business model

Take advantage of strength in broadcasting. When compared to terrestrial competitors, a major strength of satellites is in broadcasting, a form of transmission that is particularly well suited for the delivery of content directly to the home of consumers. Satellite actors should take advantage of this strength by concentrating on the provision of services that can rely on datacasting and require no, or only a limited amount of, interaction. In this way, they can take advantage of the large base of existing DTH subscribers.

Take advantage of capacity to tap underserved markets quickly. Another strength of satellites is their ability to provide service quickly in underserved

Figure A.3. **The broadband entertainment value chain**

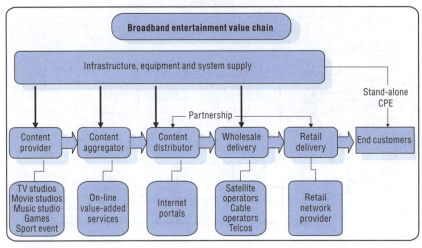

Source: Authors.

markets where competing technologies are not yet present. This should be fully exploited, notably in emerging markets where satellite entertainment could make significant inroads (*e.g.* Central and Eastern Europe, Middle East, Latin America, Asia-Pacific).

Take advantage of complementary technology. DTH operators can "save on two-way communications" (where they are weak compared to terrestrial competitors) by "pushing" content directly into the home of consumers, taking advantage of the major progress achieved in storage capacity of personal video recorders (PVR). As most of the interaction takes place between the consumer and the set-top box/PVR, traffic on the return channel is significantly reduced.

Forge alliances with others. In the longer run, the ability to deliver compelling content to consumers at a reasonable price will be crucial to success in the market place. This means that DTH platform operators should pay particular attention to the forging of strong relationships with content producers and aggregators. Alliances with telecommunications companies may also be effective in markets where competing digital cable operators offer "triple play" (*i.e.* TV, telephony and Internet).

Success factors in the business environment

Availability of the Ka-band. While other bands are crowded, the Ka-band has remained largely open. It provides ample room for the development of new satellite-based services, notably broadband and HDTV services.

Development of standards. The adoption of open standards has greatly enhanced the cost effectiveness of DTH services. It will be also a main factor for the success of satellite broadband. In particular, it can help to reduce significantly the cost of equipment by promoting interoperability and by facilitating the scalability of networks. Adoption of "plug and play" standards for the equipment installed in the home of the final consumer (customer premises equipment – CPE) should also greatly increase the attractiveness of such equipment, including PVRs.

Digital divide concerns. The desire of governments to reduce the digital divide between urban and rural areas induces policy makers to pay more attention to solutions – including satellite broadband – that can effectively deliver services in rural areas.

R&D in satellite broadband technology. Efforts under way should contribute to the introduction of new technologies that provide for more efficient use of spectrum and cut dramatically the cost of delivering satellite broadband.

Regulation. Regulation has traditionally tended to favour incumbents and existing services and to slow the introduction of innovative services. Changes in the regulatory regime (e.g. introduction of spectrum trading) may also be very disruptive to the business model of operators.

Competition from terrestrial technology. Progress in terrestrial wired and wireless technologies may gradually erode the satellite broadband niche market in rural and remote areas (e.g. ADSL, WIMAX). In the longer run, satellite entertainment as a whole may be vulnerable to the extension of fibre optic networks.

Market evolution. Two trends might adversely affect satellite entertainment: i) the trend towards "triple play": consumers may be tempted in future to prefer to deal with one supplier; and ii) changing consumption patterns, i.e. the growing demand for bandwidth and interactivity might put satellite at a disadvantage (the Korean model). The Korean experience is interesting because Korea is ahead of the rest of the world in terms of broadband access. More than half the population has had broadband access for more than three years. Particularly significant has been a 650% rise in online gaming from 1999 to 2001. As a result, the market share of online gaming has risen from 2% to 21% of the overall gaming market. If online gaming becomes as popular in other countries, as the penetration of broadband increases, the Korean experience could signal how consumer entertainment will change throughout the world. Gaming may presage changes in the value chain that will become common in other areas.

Economic/business models

Consideration of the critical factors outlined above suggests a strategy for the development of satellite entertainment. It involves the adoption of three business models:

Model 1: HDTV. By fully exploiting the current strength of satellites in the provision of broadcast services and the large existing subscriber base, this model involves the aggressive development of HDTV: it is almost non-existent in Europe today and emerging in the US market. HDTV offers higher-quality picture and sound than regular TV. Market studies have demonstrated that there is a significant pent-up demand for higher-quality images among consumers, notably for films and sporting events. It also involves the development of digital cinema: as content is increasingly produced in digital form, digital cinema represents a natural extension that satellites could fully exploit. The model would involve a partnership between satellite operators, HDTV distributors (to households as well as to digital cinemas), digital video broadcast content producers and manufacturers of high-definition equipment.

Model 2: ITV. This models develops iTV services, by taking advantage of the capability of the new PVRs: such devices offer increasing opportunities for providing quasi-interactive services via satellite, including video-on-demand, as their cost declines and as their capability and ease of use improve. For the DTH platform, the business strategy is first to expand the base of equipped subscribers by marketing aggressively services offering operator-sponsored PVRs, boosting in this way subscription revenues, as PVR clients generate more revenue than non-PVR clients. Once a critical base of customers is created, new possibilities for raising advertising revenues arise because the technology offers advertisers new opportunities to target their advertising messages more effectively.

Model 3: Satellite broadband. This model takes advantage of access policies to deploy broadband in rural and remote areas and introduce new technologies to cut costs. Satellite is the only technology able to fully bridge the digital divide. With public support, satellite broadband could be deployed in rural and remote areas where market demand is inherently weak. The use of the Ka band and new technologies (*e.g.* use of spot beams) that conform to newly established open standards (*e.g.* DVB-S2) could contribute to significant efficiency gains. A credible business model for satellite broadband entertainment needs to be based on competitive subscription rates, cheap terminals with very large storage capability and very sophisticated and easy to use electronic programme guides, WIFI connection for all the equipment installed in the home of the final consumer, and the provision of compelling entertainment content. This requires a strong partnership of several players in the value chain, including the manufacturer of equipment, the distributor of content and the operator of the broadband network.

Prospects for the future

Model 1. Several factors will contribute to the rapid development of HDTV in the coming years, including the rapid decline in the cost of HDTV equipment and display as well as its attractiveness on the larger screens that will become the norm in future. DTH platforms are well-placed to take advantage of this development because of their strength in broadcasting and their large base of satisfied customers. On the negative side, limitations on the availability of frequencies might be a problem as HDTV uses twice as much spectrum than regular TV, even with progress in compression. The success of satellite HDTV is likely to be greater in a more open environment (the Smooth Sailing scenario) where the cost of equipment declines faster and where the cost of production of content can be spread over wider geographical regions than under more restrictive conditions (the Back to the Future and Stormy Weather scenarios).

Model 2. Like HDTV, iTV appears to have a bright future, not only because of progress in the technology (*e.g.* PVRs, electronic programming guides, or EPG) but also because of viewers' desire to have more control over their viewing experience and the desire of television announcers to target their commercials more effectively. DTH platforms should be able to capture a fair share of this new market because of their inherent strength in broadcasting and their large subscription base. Moreover, progress in satellite broadcasting technology (*e.g.* implementation of DVB-S2) should make the delivery of satellite iTV more cost-effective. However DTH operators will need to face growing competition from digital cable operators. As for HDTV, the development of satellite iTV is more promising under the more open environment of the Smooth Sailing scenario than under the other two.

Model 3. Broadband entertainment is likely to expand rapidly in the future as an increasing share of the world population has access to the broadband network. Given the inherent weakness of satellites for two-way communication, satellite broadband entertainment will largely be restricted to rural and remote areas. However, progress in satellite technology should contribute to bring costs down to more competitive levels. Moreover, strengths of satellites in broadcasting may continue to give it an edge over its rivals: entertainment is typically produced for large audiences rather than tailor-made for each individual, because of the large fixed cost in content production and the economies of scale in displaying content. Hence, datacasting is likely to remain the most efficient technique for bringing such content to users, including those on the move in an increasingly mobile society. Once again, the open environment of the Smooth Sailing scenario is more favourable to satellite broadband entertainment than the Back to the Future and Stormy Weather scenarios.

ANNEX A

3. Location-based Service Applications: Road Traffic Management

Introduction

While road transport brings major benefits overall, it also imposes substantial costs on society at large. For instance, the economic costs of congestion and road accidents may amount to a staggering EUR 400 billion in Europe alone. Worldwide, road transport produces roughly 20% of carbon emissions and smaller shares of the other five greenhouse gases covered by the Kyoto Protocol (IEA, 2002). The situation may worsen in future. By 2030, the number of light vehicles and of kilometres travelled may more than double in the world (+137% for both), with an even larger increase for heavy vehicles (+190% vehicles and +192% kilometres travelled). By 2030, more than 1.5 billion light vehicles and 87 million heavy vehicles may be on the roads (OECD, 2001).

More effective traffic management should result in more efficient and safer use of transport infrastructure. This will translate into significant savings in terms of reduced costs of accidents (from a human as well as economic point of view) and of congestion, as well as a lesser need for the construction of new infrastructure.

Location-based services (LBS) can make a difference in this regard. Broadly speaking, they rely on the ability to locate an individual or an object (*e.g.* a car) in real time, using either satellite technology – notably global navigation satellite systems (GNSS) – or terrestrial technology or both. They have a broad range of applications, including transport telematics applications (*i.e.* the blending of computers and telecommunications, which combines wireless communication with location technologies such as GNSS to provide services to vehicles on the road) and the management of road traffic. Over the coming decades, such applications may help alleviate road traffic problems in terms of pollution, congestion and accidents.

The main actors

Several actors are involved in the provision of LBS. They include first of all customers, who may be public (*e.g.* road transport authorities that use LBS to monitor and control traffic) or private (*e.g.* road carriers that use LBS for fleet management, or individual users who acquire telematics equipment or subscribe to telematics services). They also include providers of navigation signals; providers of chipsets (*e.g.* GPS chips) and navigation equipment; the car industry (which integrates navigation equipment such as vehicle protection systems in new vehicles and other location-based telematics equipment); and the insurance industry (*e.g.* introduction of "pay-as-you-drive" insurance policies for which the premium is based on usage which can be monitored by location-based devices).

Figure A.4 illustrates a situation in which the signal provider consists of a system operator created in the context of a public-private partnership (PPP) that offers three signals: a free signal available to all, an enhanced public signal restricted to designated public users, and an enhanced commercial signal. The operator has three main sources of revenue: i) payment for the use of the enhanced public signal by the government, which can take the form of a fixed guaranteed level of revenue for such services; ii) royalty payments by the producers of the chipsets; and iii) service fees paid by providers of value-added services that use the commercial signal.[1] The car manufacturing industry appears in Figure A.4 not only as a customer of navigation equipment subsystems, but also as a provider of such systems when selling vehicles to the mass market. Moreover, car manufacturers may set up subsidiaries to offer telematics services to their clients.

Figure A.4. **General system mapping for LBS and road management**

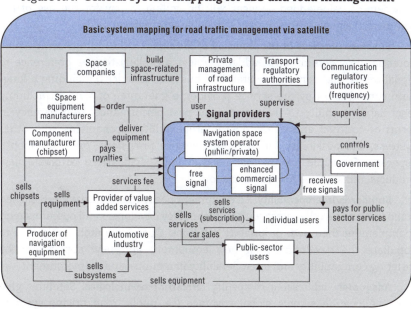

Source: Authors.

ANNEX A

Critical success factors

Technology. The signal provided must meet users' requirements in terms of accuracy, availability and integrity. This means that several signals may have to be on offer to meet these needs. The quality of the signal can also be enhanced by signal augmentation systems, either space-based (*e.g.* the European EGNOS) or ground-based. In future, interoperability between navigation systems (*e.g.* between GPS and Galileo) will enhance the quality of available signals.

Revenue streams. These need to be clearly established (public payments or royalties on chips or service fees) and sufficient to cover the large expenditure needed to acquire or upgrade the physical systems and the operating expenses, and provide an adequate return on investment.

Liability. The operators' risks should be clearly identified and the extent of their liability established, so that such risks can be insured.

Pricing. The critical success factor here is not to alienate the users with overpriced services. The GPS experience suggests that there is a rapidly growing demand for onboard navigation equipment based on the free GPS signal, where the GPS chip represents a very small cost item.[2] If a modest royalty was to be set on such chips (or the price of chips used by other navigation systems), the impact on the price of the equipment, hence on demand, should be very minor.[3] On the other hand, there is more uncertainty regarding the service fees providers of value-added services will be willing to pay for the use of the commercial signal, as this market remain largely undeveloped.

Business practices in the provision of the signal. Other things being equal, potential providers of value-added services are likely to give preference to signal offers that conform to standard business practices, including a long-term commitment by the service provider; a guarantee of service, including legal liability; ability to trace and audit past performance; transparency in terms of contract and operations; interoperability mechanisms with other systems; the possibility for integrated service provision ("one-stop shop"), especially for value adders that need signal specifications to create new products and services for road transport.

Standards/interoperability. All equipment needs to meet clearly established standards and be fully interoperable. Technology developers and operators tend to implement their own proprietary solutions. Open standards would help to stimulate the development of new equipment and new applications by fostering compatibility and interoperability in a multi-vendor environment.

Adequately addressing competing terrestrial technologies. Satellites give users of location signals global coverage (*i.e.* public road pricing systems developed over large areas, international road transport companies), although there may be an increasing merging of terrestrial communication technologies that compete directly with satellites. It is therefore important to ensure that space-based systems continue to complement usefully the large interconnected terrestrial-based communications networks to come.

Positive attitude of the auto and insurance industry. These are important stakeholders in the overall system as technology developers, customers of specific navigation products, and service providers to the mass market. Their influence needs to be taken into account for road transport applications to succeed.

Governments. The favourable political environment for improving road management should be reflected in decisive actions by governments, notably by departments of transport, to promote the development and implementation of road management systems. Road pricing should not only help to reduce pollution and congestion, it could also be a useful source of additional revenue for improving the transport infrastructure.

Economic/business models

Model 1: Car navigation. This model provides, for a fee, in-car telematics services that take advantage of the capacity to locate the vehicle in real time. Such services include roadside assistance, automatic notification of airbag deployment, tracking of a stolen vehicle, guidance for lost drivers, traffic information, dynamic re-routing and personalised services, including news, stock quotes, weather, messaging. The service may be offered by a subsidiary of the car manufacturer (*e.g.* OnStar, a subsidiary of GM). It could also result from a strategic alliance between two complementary players (*e.g.* the alliance between Ford and Vodaphone in Europe to offer telematics services involving a combination of GSM and GPS technologies).

Model 2: Road user charging. In such a model, a LBS device in the car provides information electronically to the road toll centre on distance travelled, route used and time of travel. Such information is then used to charge the user's account. The user is presented regularly with a bill, like any other utility bill, such as that for electricity or water.

Prospects for the future

Model 1: Car navigation. The "intelligent car" concept is gaining ground as advanced plug-in satellite navigation sensors are increasingly sold to original equipment manufacturers (OEM) that are developing applications for the automotive industry (Garmin, 2003), and partnerships between car manufacturers and communications operators are being set up to provide location services (*e.g.* Ford and Vodaphone).

This trend may accelerate in future owing to: *i)* the success of road assistance insurance schemes and the desire for increased safety; *ii)* the aspiration of users to reach their destination faster and more conveniently; *iii)* users' need to have access to a broad range of services when in their car; and *iv)* interest from insurance companies if such devices help to reduce claims.

Major catalysts could be the explosion of smart mobile devices (*e.g.* mobile phones equipped with GPS) and the development of satellite radio and broadband wireless as a way of downloading content to the vehicle's digital entertainment system.

The development of car navigation is likely to be faster in the open environment of the Smooth Sailing scenario than in the other two because it is in this scenario that mobility is likely to be the greatest, that the cost of equipment declines the most and that space-based solutions are the most effective, notably through the interoperability of GNSS. Prospects for car navigation services are still good under the Back to the Future scenario, although costs are likely to be higher, as economies of scale cannot be as fully exploited. In the Stormy Weather scenario, development takes place more slowly as populations are poorer and less mobile, costs remain high, while lack of interoperability and standardisation across countries limits the value of such systems for users.

Model 2: Road user charging. There are already several road charging demonstration and pilot projects using satellites under way worldwide, notably in Europe where there is great interest in the development of such systems. For instance, the European Commission proposed in July 2003 a new directive on road infrastructure charging which will, from 2008, allow for differentiating road tolls according to the type of vehicle and infrastructure, as well as the time, period, location and distance travelled, taking environmental and accident costs into account in the pricing. In future, such schemes are likely to be more widely implemented as technology improves and traffic congestion increases.

The impact of road charging could extend beyond car use and traffic flows. In particular, it could have far-reaching effects on the geographical distribution of activities (*i.e.* with road pricing, it might be cheaper to buy in the local store than in the distant shopping mall, thereby encouraging a renaissance of local trade).

The development of road user charging appears inevitable in all scenarios. Although the level of mobility is higher in the Smooth Sailing scenario than in the Back to the Future scenario, where it is still higher than in the Stormy Weather scenario, the level of public and private resources available for developing the road infrastructure varies as well and in the same direction. Hence, in all scenarios there will be growing pressures to curb traffic growth, and thus to apply road user charging schemes.

Notes

1. For instance, in the business model developed by PWC for Galileo, it is estimated that the royalties on chipsets should represent the principal source of revenue in the early years of the system, while service revenues should develop over time as the market evolves (PWC, 2001).
2. The price of GPS chips is currently around USD 10 and is declining by 15% a year.
3. In the case of Galileo, PWC (2001) has estimated that the royalty on chipsets should not exceed EUR 0.50 per chipset. The study further notes that the proposed royalty level could be one of the criteria for bid evaluation in the final concession bid.

4. Risk and Disaster Management

Introduction

During the last decade, natural disasters affected more than a billion people and assets worth approximately USD 730 billion were destroyed, according to the *World Disaster Report* (International Federation of Red Cross and Red Crescent Societies, 2002). Moreover, the losses from such disasters are increasing, especially in less developed countries.

In an international context where development of effective risk and disaster management systems is becoming a major concern for governments (OECD, 2003a), there is growing interest in the capabilities Earth observation (EO) may offer for improving their effectiveness. Aside from weather satellites, which have proved their worth in the monitoring of dangerous weather conditions (WMO, 2004), attention is increasingly direct to the contribution that other EO satellites might make at different stages of the risk and disaster management (RDM) cycle.

RDM reflects the way society is organised to deal with hazards. In this context, a *hazard* is defined as a potential source of harm, *i.e.* a phenomenon that may cause disruption to humans and their infrastructure. It can be natural (*e.g.* fire caused by lightning) or man-made (*e.g.* accidental or voluntary fire). A *disaster* is an event that causes such disruption. The term *risk* refers here to the likelihood that a hazard will cause a disaster. Hence, risk and disaster management is a set of actions and processes designed to minimise the effects of a disaster (CEOS, 2003).

Several types of actions can be taken in RDM. These include, first, risk management measures, *i.e.* measures to reduce the risk of occurrence of the hazard in the first place (*e.g.* forbidding smoking in gas stations) and preventive measures to reduce the risk of extensive damage if the hazard occurs (*e.g.* use of fire-resistant material in fire-prone areas). Second, RDM involves disaster management measures, *i.e.* measures designed to ensure quick and effective response when the hazard occurs (*e.g.* having a fire extinguisher at hand and be trained to use it) as well as rapid and effective recovery after the disaster (*e.g.* having well-trained and well-equipped first responders arrive quickly to the site of the disaster and take care of victims effectively).

ANNEX A

More generally, RDM involves three main phases (mitigation, emergency management, recovery), as illustrated in Figure A.5.

Figure A.5. **Risk and disaster management cycle**

[Diagram showing the Risk and Disaster Management Cycle with phases: 1. Mitigation (Risk assessment, Preparedness, Planning, Prevention), 2. Emergency management (Disaster, Response), 3. Recovery (Rehabilitation, Reconstruction)]

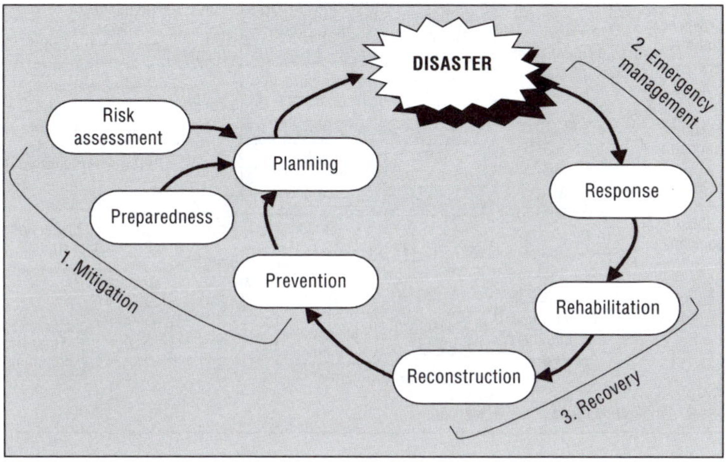

Source: Adapted from World Health Organisation (2002), J. Adams and B. Wisner (eds.), *Environmental Health in Emergencies and Disasters: A Practical Guide*, World Health Organisation, Geneva.

An effective RDM system requires the timely input of actionable data and information in all three main phases of the management cycle. EO data may be one such input. The following is an exploration of how it could be applied to RDM in the case of two specific hazards (namely fire and flood). The examples serve to explore some of the key features of possible economic/business models that might be used for this purpose, identifying in the process the factors that are essential for their effective implementation.

The main actors

The provision of space-based data for RDM may involve a large number of actors on both the supply and the demand side. Several configurations may be envisaged on the supply side:

- The provider may own and operate EO satellites dedicated or not to a particular type of hazard. The data can be directly provided to users or processed in house before delivery to RDM users.
- The provider may acquire the image from an EO satellite operator as well as other sources and process them for various RDM users.

Figure A.6 illustrates a situation in which the provider of EO data for RDM is a combination of public actors (*e.g.* space agencies, specialised application agencies) and private actors that operate EO systems for EO data acquisition,

ANNEX A

Figure A.6. **The systemic view for risk and disaster management**

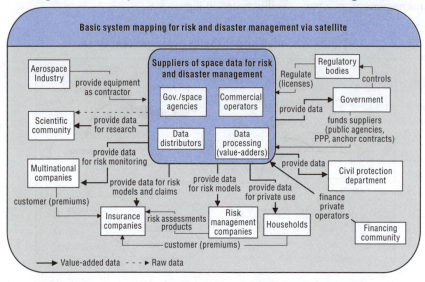

Source: Authors.

as well as essentially private actors that distribute and process data. Taken as a group, these actors provide both raw data and value-added data to public and private users.

Public users include civil protection departments as well as other government agencies that may need such data (*e.g.* environment agencies, departments of agriculture, commerce, tourism, transport). Private users may include insurance companies, multinational enterprises (MNEs), risk management companies, civil aviation, the media and even households (*e.g.* hazard maps when buying real estate). The scientific community is another important user of EO data for RDM and typically might use both raw data for some research applications as well as value-added data.

Other actors depicted in Figure A.6 include the regulatory bodies that issue licences to commercial operators and regulate their activities (*e.g.* in the United States, NOAA acts both as a EO data provider and as a regulator of private EO systems operators), the financial community, which may invest in private providers of EO data, and the aerospace industry, which provides the equipment necessary for the development of the EO systems.

Other actors might also be involved, such as suppliers of the hardware and software needed for data processing, data distribution and data archiving or the providers of broadband communication services used for distributing the value-added data to final users.

Critical success factors

The nature of the service. The nature of the service provided very much depends on the use to be made of the data collected by the system. Typically, high revisit frequency, flexibility and speed in tasking, and high resolution will be required in the emergency phase of the RDM cycle. However, the requirements may be quite different in other phases. For instance, in the mitigation phase, the main need for EO data will be for the creation of hazard maps and inputs to predictive models. This will require high-resolution data but not high revisit frequency. During the rehabilitation phase, high-resolution data will also typically be needed for the planning and monitoring of the relief effort. The space system must also be suitable for the mission at hand. For instance, when heavy clouds or smoke covers the hazard scene, optical satellites will not be able to provide useful EO data, so that radar satellites would have to be used instead.

How the service is to be provided. On the production side, an important consideration is whether to adopt a multi-purpose or a dedicated approach. If conventional remote-sensing satellites are to be used, it is imperative for such spacecraft to fulfil as many requirements as possible so as to spread the fixed cost of the space asset over a maximum number of possible applications. Therefore, the mission specifications (particularly the spectral bands) should be chosen to provide good general-purpose imagery, suitable for a wide range of applications.

On the other hand, if micro-satellites can be considered, the economics of the system are significantly altered. Micro-satellites are sufficiently inexpensive to be affordable to individual government agencies or companies for their specific remote-sensing tasks. The spectral bands (or resolution, or orbit) of a dedicated imaging micro-satellite can be optimised for specialist requirements, which are not met effectively by general-purpose spacecraft. Another advantage of micro-satellites is that, because they can be deployed cheaply and quickly, they can be used to test new technologies within a short time frame. Constellations of micro satellites also offer greater continuity of service than conventional satellites.

How will revenue be generated? Sufficient revenue to support the activity is obviously a major preoccupation of EO operators, whether public or private. The public model is viable only if the public financing is fully secure on an ongoing basis. This is not always the case however. For instance, space agencies that develop demonstration EO satellites are typically not in a position to continue to finance the system beyond the demonstration stage, so that it is important to create a sound alternative source of financing for such systems. A private operator needs to ensure that the revenues received from government contracts and commercial clients are sufficient to cover both capital and operating expenses. So far, government contracts have represented the lion's share of revenues since the commercial market has not developed as originally hoped.

Open data standards. Open standards for the database that would integrate data from various sources would greatly facilitate the use and reduce the cost of using such data and provide as well opportunities for developing new hardware and software.

Level playing field. The development of the sector, notably the entry of private actors, would ideally require that all EO data providers be treated equally, so as to foster competition. However, this is not likely to occur in practice, given the importance of security considerations and the role played by public actors. Moreover, markets for EO data are so narrow that, in most cases (the United States is an exception), there may only be room for one firm in any given market. Even when the market is large enough to accommodate several players, firms may need a guaranteed revenue stream from the government (anchor tenancy contracts), as government is the main if not the only client for EO data.

Policies supporting the RDM process. Governments have a major responsibility to put in place policies that are supportive of RDM, such as policies that encourage municipalities to take stock of existing infrastructure, assess their vulnerability and develop contingency plans or civil security policies for strengthening their ability to anticipate disasters and respond effectively. Efforts to create RDM systems at international level are also important, notably for meeting the needs of less developed countries.

Involvement of users. Greater awareness of users and their implication in the definition of data requirements would give producers of EO data a firmer basis for making appropriate business decisions. The production and distribution of EO data must be fully integrated in the RDM cycle.

Economic/business models

For two important hazards (fire and flood), selected elements of the economic models for developing EO operational monitoring systems are discussed: the services that need to be provided, the main users to be served and the infrastructure needed to provide such services. None of the systems available today is fully operational.

Model 1: Fire management. An effective EO system needs to be capable of supporting effective disaster warning, response and recovery and of generating information products that enable planning and mitigation. This consists first in developing information on the nature and status of the biomass (volume and moisture), as well as EO data that can be used for assessing the possible impact of a fire (taking land-use patterns into account) and for taking preventive measures (*e.g.* clearing bush near built-up areas) and preparing emergency measures (*e.g.* access and escape routes).

The EO system should also provide data for fire detection (*i.e.* for detecting the position of the fire, its size and intensity, presence of fuel, and conditions) and monitoring (*i.e.* providing geo-reference images of the fire front with information on the surroundings). Finally, immediately after a fire, EO data should be made available for assessing the amount, degree and location of damage to structures and infrastructure and the condition of the soil and vegetation resources.

Main users. The main users of EO data for fire hazards are a very heterogeneous group with different requirements. They include RDM services on the ground (*e.g.* fire prevention personnel, emergency preparedness personnel, civil protection agencies, relief agencies), ministries with a mandate in this area (ranging from interior, agriculture and public works to national environment and public health agencies), international organisations (*e.g.* UN Environment Programme, International Panel on Climate Change, UN Food and Agriculture Organisation), the scientific community (including space agencies) as well as private-sector actors (*e.g.* land managers, news media, aviation communities, insurance companies, transport planners). Serving these diverse needs is a major challenge for the providers of EO data.

Infrastructure needs. Today, there is no operational system for monitoring fires. The various current fire monitoring activities are largely the domain of research. The MODIS (moderate resolution imaging spectroradiometer), a key instrument aboard NASA's Terra and Aqua satellites, and DLR's bi-spectral infrared detection (BIRD), the world's first dedicated small satellite for fire recognition, provide perhaps the most interesting prototypes of future operational fire monitoring systems.

To improve operational use of the available information, more attention needs to be given to data availability, product accuracy, data continuity, data access and how the data are being used to provide useful information. There is currently no standard *in situ* measurement and reporting system, and national reporting is extremely variable and wholly inadequate to provide a consistent regional or global assessment. It is also often hard to relate satellite and *in situ* data reporting. In the next few years it will be necessary to develop not only appropriate standard methods for fire monitoring but also the institutional infrastructures for operational global fire monitoring and reporting.

Model 2: Flood management. EO data can be used at all stages of the flood RDM cycle. First, they can be used as input to the mapping of geomorphic elements and land use, providing meteorological data for hydrological modelling and contributing to mapping historical events. Remote sensing may also be used to map topography and define surface roughness and land cover as well as to update cartography for land use. Such data are used to develop the hydrological models that play a major role in assessing and forecasting flood risk. Model predictions of the potential extent of flooding can help emergency managers develop contingency plans well in advance of an event to help facilitate a more efficient and effective response.

During a crisis, remote sensing is the only cost-effective method of monitoring the spatial extent of flooding and is essential in areas that are not instrumentally monitored. EO data can also be used in this phase to make a first assessment of damage, including damage to buildings and infrastructure, and to evaluate secondary disasters, such as waste pollution.

During recovery, a more detailed assessment of the damage can be made, using both medium- and high-resolution remote-sensing images, together with an operational geographic information system (GIS). The medium-resolution data can establish the extent of the flood damage and can be used to establish new flood boundaries. They can also locate landslides and pollution due to discharge and sediments. High-resolution data is suitable for pinpointing the location and degree of damage. They can also be used as reference maps to rebuild bridges, washed-out roads, homes and facilities. Finally, they can be used by agencies to validate and refine the hydrological models that are used for flood prediction.

Main users. Users and customers of EO products vary over the RDM cycle. In terms of prevention, they include land planners (federal or national), hydro-meteorologists (including weather forecasters) and environmental and agricultural authorities. At the mitigation and prediction phase, the main users are emergency managers. In the preparedness warning phase (immediately before the flood), the users include civil protection specialists, hydro-meteorologists, local authorities, water management and the media. In the response phase (i.e. during the flood), users are essentially the same as in the preparedness category and also include insurance companies. In the recovery phase, the data are used mainly by land planners (federal or national), hydro-meteorologists (including weather forecasters), and environmental and agricultural authorities, and insurance companies

Infrastructure needs. Satellite data for flood management are provided by polar orbital Earth resource satellites and operational meteorological satellites. The former are of two types: i) optical sensors that cannot see through clouds and operate at low, medium and high resolution; and ii) microwave sensors that can see through clouds and include high-resolution active sensors such as synthetic aperture radar (SAR) and low resolution passive sensors (SSMI).

Meteorological satellites are also of two types: geostationary and polar orbital. The geostationary satellite (GOES) is a powerful tool to observe the weather on a continuous basis. The orbit is at an altitude of 22 000 miles and picture frequency is normally on a half-hourly basis. Polar orbital satellites (POES) circle the earth twice a day at an altitude of approximately 850 km. Both of these satellites provide visible light and infra-red imagery and microwave data. With respect to precipitation estimates and moisture analysis, GOES offers higher resolution time and space scales, while POES microwave data are more physically related to precipitation and moisture processes.

Prospects for the future

The future of EO systems for RDM will very much depend on the ability of such systems to fully meet users' data needs in terms of accuracy, spatial and temporal resolution, timeliness and geographical coverage. Moreover, EO data require extensive integration in diverse data streams to be used for the generation and dissemination of the timely and accurate information needed by decision makers and the public. Such information includes, for instance, forecasts of the likely impact of the hazard event (e.g. flood) that is generated by predictive models that use EO data as one of their inputs. In the case of floods, EO provide data on the volume of rain expected to fall in a particular region, which can then be entered in a runoff model to assess its impact. The construction of the runoff model itself uses EO data (e.g. phase information obtained from SAR interferometry is used to derive digital elevation models, a major input to runoff models)

Although progress has been made in recent years in terms of data acquisition and predictive modelling, the requirements are not yet fully satisfied. Indeed, weather apart, few of the observational requirements related to major hazards are now available on a worldwide basis.

For instance, an important data gap is the lack of a suitable measure of soil moisture. Measuring soil moisture by satellite is still at the research state and may not be available on an acceptable operational basis for another ten years. This is very unfortunate, since soil moisture plays a key role for assessing the impact of a number of hazards, including fire and floods.

Moreover, some of the basic background information needed for generating meaningful hazard zoning maps, which are essential for planning and mitigation efforts, are inadequate to support disaster reduction strategies. Finally, co-ordination between observation organisations and research communities remains weak. EO information – whether from space, airborne or ground-based systems – is not used consistently in disaster management decision-making.

In future, some of these problems should gradually be addressed. First, new generations of satellites that take advantage of progress in satellite sensor technologies should contribute significantly to reducing existing technical gaps. In particular, progress in nanotechnologies and in optoelectronics and advances in onboard processing should greatly increase the capabilities and cost-effectiveness of space instruments (OECD, 2003b). For instance, very effective neuronal network processors show very good progress in machine intelligence and autonomy for aerospace systems. Such progress could be very valuable for disaster warning and hazard detection, which require quick classification and a short response time. Such tasks will require implementing a high-level data processing chain onboard the satellite.

Second, efforts under way to facilitate the distribution of EO data and their effective integration in GIS should encourage further use. Advances in software, data compression and archiving, coupled with the generalisation of broadband access, should offer opportunities for increasingly sophisticated user-friendly geospatial systems. Lower engineering costs are making direct broadcasting more easily achievable. This, in turn, enhances the opportunity for the development of real-time EO applications. The use of broadband Internet access is seen as an important factor in providing immediate services to private customers (*e.g.* urban planners, insurance companies).

As hardware and geospatial software become more affordable and more widely available, the prices of some standard products for disaster management should decline, especially with the development by the GIS industry of open standards.

Another strong driver that will foster the development of EO systems is the keen political interest in EO data in general, on sovereignty and military security grounds. There is also a strong interest in EO data in the scientific community. Hence, outside the military and scientific spheres, EO can be seen as a technology looking for applications beyond the well-established weather prediction. In this regard, the use of EO for RDM is attractive. It is a way to recognise the need to extend the concept of security more broadly to the civil sphere. Moreover, applying EO to RDM could be an effective way to project "soft power" at international level for two main reasons. First, RDM is an activity for which there is substantial scope for international co-operation. Second, EO may be viewed as an ideal tool for this purpose, given the ubiquity of space-based systems and given that, in many parts of the developing world, no effective alternative exist for generating the data necessary for RDM.

But meeting users' data needs is not enough. EO systems will only be used for RDM if the benefits are seen as substantial. However, such benefits are largely at the margin, since EO data represent a small input to the overall RDM process. In the case of fire, for instance, the most significant benefits are likely to be in terms of cost avoidance. This would be the case if EO data allow for the implementation of more effective fire-prevention measures as well as faster detection of fire and more rapid suppression. Benefits could arise also for more rapid recovery and avoidance of harmful secondary effects (*e.g.* faster and more precise identification of burnt areas makes possible faster and more effective implementation of measures designed to prevent landslides and to restore vegetation).

The same applies to flooding. Even if an ideal EO system were in place, the EO data produced would only be one of the inputs into the flood RDM process and could only improve flood management at the margin. In the short term, improvements would result largely from more effective preventive measures, and in the longer term, from changes in land use, limited soil sealing, reforestation, encouraging urbanisation in less risky areas, etc.

However, even a marginal improvement in the process could generate major economic and social benefits, given the high costs of flooding (for instance, the United States suffers an average of 225 casualties and more than USD 3.5 billion worth of property damage from flooding and heavy rainfalls every year). Hence, even if better use of EO data contributed to reduce damage by only 1%, this would represent for the United States alone a saving of the order of USD 35 million a year.

When considering the different visions of the future depicted in the scenarios developed in the second phase of the project, it is clear that the application of EO data to RDM is likely to be more extensive in the Smooth Sailing scenario. It offers the best prospects for international co-operation and the cost of the systems is likely to be lowest, with opportunities for the development of global dedicated systems using constellations of micro-satellites. Significant progress is still likely to be achieved in the Back to the Future scenario because security concerns are greater while the scope for effective international co-operation, although less than in Smooth Sailing, is still substantial. Prospects are less favourable in the Stormy Weather scenario, as international co-operation is severely curtailed, resulting in substantial duplication of effort across countries, and the cost of equipment remains high as markets remain fragmented.

5. Space Tourism

Introduction

In general terms, space tourism can be described as the business of taking ordinary healthy citizens who can afford it to space without special training. It may well eventually be a segment of the fast-growing tourism/adventure sector. Not only is there interest among the general public in such a venture, but progress in sub-orbital space travel may put a much more affordable way to reach the frontier of space within reach. Adventure tourism already accounts for a world market of around USD 50 billion a year, and space tourism has the potential to become one of the largest commercial space applications if a credible and affordable offer can be put together. Another reason for selecting this particular application is that space tourism may contribute, over time, to a reduction in the cost of access to space.

Space tourism operators might provide two different types of space travel: suborbital flights and orbital flights. In suborbital travel, customers are flown to space for a very short time on a parabolic trajectory culminating above a threshold altitude of 100 km, considered the edge of space (although not officially recognised as such by international space law). The entire suborbital experience only lasts a few minutes. It is estimated that from take-off to landing, flights may last about 15 minutes, including a few minutes of weightlessness. In orbital travel, customers are flown to space using an expendable or reusable launch vehicle able to reach orbital speed (about 8 km/s) and possibly to attach to an orbiting station like the International Space Station (ISS) with living quarters. The customers are then able to stay for a few days or weeks in the orbital facility.

The main actors

Many different actors have a role to play in the development of space tourism. They include space tourism companies and their shareholders (aerospace companies, travel and/or tourist companies, media companies, investment funds, banks, individual shareholders); potential customers (healthy citizens interested in travelling to space and able to fund their space venture directly or from collective support provided by sponsors, media [reality shows, games], lotteries, etc.); the aerospace industry which conducts the development and manufacturing of spaceships or space-planes and related ground equipment and facilities; the tourism industry which carries out the marketing and sales and provides additional services (lodging for customers and families, etc.); and space agencies, which can provide support for the development of space-planes suitable for space tourism and related research.[1]

Figure A.7 illustrates how the different actors are configured in the case of a private company, Space Tourism Co., which provides suborbital packages.

Figure A.7. **System mapping for space tourism**

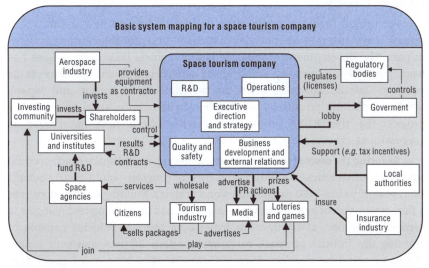

Source: Authors.

Critical success factors

The development of space tourism will hinge on a number of factors that will play a critical role.

Overcoming the legal and regulatory hurdles. The full development of space tourism depends crucially on the resolution of a number of legal and regulatory issues related notably to whether space tourism should be regulated like the airline business and what the space tourism company's liability towards customers should be: Can customers be asked, after having had the risks clearly explained to them, to accept the risks without any recourse against the space tourism operators or other involved parties? Space tourism accidents may be much more visible than long-distance sailing or general aviation accidents and may negatively affect the business prospects of all space travel ventures. Liability issues also arise with respect to third parties: What level of third-party liability could private space tourism start-ups bear and would government act to limit this liability? Finally, it still needs to be settled how spaceports should be regulated, particularly from the point of view of environmental acceptance.

Developing the right technological solutions. A sharp distinction needs to be made between suborbital and orbital flight. For suborbital flight, no major technological hurdle has to be overcome. Technologies and subsystems are available off the shelf and simply need to be integrated and tested as tourist transport systems. The main challenge will be operational rather than technical, *i.e.* to operate suborbital space-planes so that they fly often with little maintenance between flights, a challenge familiar to aviation but not to the space industry.

For orbital flights, the challenge will be to develop new orbital space transport systems that are much less expensive and probably reusable, at least in part. The development of reusable launch vehicle (RLV) technologies has been pursued for years by some space agencies with limited success as budgets have so far been quite limited (notably in Europe and Japan). Moreover, shifting priorities have resulted in the cancellation of some RLV projects (*e.g.* the DC-XA project) while other efforts have not proved as successful as expected (*e.g.* NASA's X33).

Reducing the costs of access to space. While suborbital space tourism could start without technical breakthroughs, further progress would clearly benefit from governmental R&D programmes and projects. Building on the success of a first generation of suborbital space-planes based on existing technologies, a second generation of commercial hypersonic tourism vehicles capable of crossing the frontier of space for longer periods of time and over longer distances could be developed by 2015-20. By 2025-30, a third generation capable of launching payloads in low Earth orbit (LEO) for only USD 100/kg might become viable. This would open a new era for space tourism, with orbital flights available for a few hundred thousands of dollars.

Establishing a customer base. The price tag for suborbital space tourism is very high (in theory, typically USD 100 000 in the early years) compared to other forms of adventure tourism. This means that very attractive packages need to be developed to justify the price. Potential customers are the wealthy (*e.g.* Internet millionaires or billionaires), the sponsors of lotteries, or individuals who want to use space flight as a way to promote their business activities.

Securing investment. Private funding is essential for suborbital space tourism and should preferably take the form of equity rather than debt. Public funding of basic R&D by space agencies could usefully complement it. Overall, the investment needed for a typical small space tourism firm could be of the order of USD 250 million. For orbital space tourism, the level of investment would be much higher, but it is impossible to estimate at this stage because of the many unknowns linked to the technical challenges that must be overcome.

Economic/business models

A market study conducted by Futron in 2003 suggests that by 2021 the suborbital tourism sector could be worth USD 800 million in revenue a year with 15 000 passengers in the US suborbital market alone.[2] For orbital tourism, only about 60 persons a year are thought to be able to afford the USD 5 million price tag expected by 2021, which would generating USD 300 million in revenue for space tourism entrepreneurs. This market study serves as the background for the study of two business models.

Model 1: Suborbital tourism. In this model, a private company provides suborbital travel to private customers. Cash flow estimates for the business model are made on the basis of assumptions regarding: development costs (USD 100 million); cost of vehicles (USD 30 million per space-plane); number of space-planes acquired (five); number of flights per vehicle per year (100); cost of ground infrastructure (USD 50 million); number of passengers per flight (two); fixed operational costs (USD 10 million), variable operational costs (USD 0.02 million per flight-vehicle); ticket price (USD 100 000).

Corresponding to traffic of 1 000 passengers a year when all five space-planes are in service, this business model is consistent with the Futron forecast for suborbital tourism. In this model, the company's operating income or EBIT (earnings before interest and taxes) becomes positive in year 5 and the cumulated cash flow becomes positive after year 8, as illustrated in Figure A.8. This is still quite long by current investment standards in high-technology sectors. Hence, suborbital tourism may not yet be economically viable. However, it may very well become so in the coming years, as costs come down.

Figure A.8. **Cash flow generated by the space tourism company**
Millions of EUR

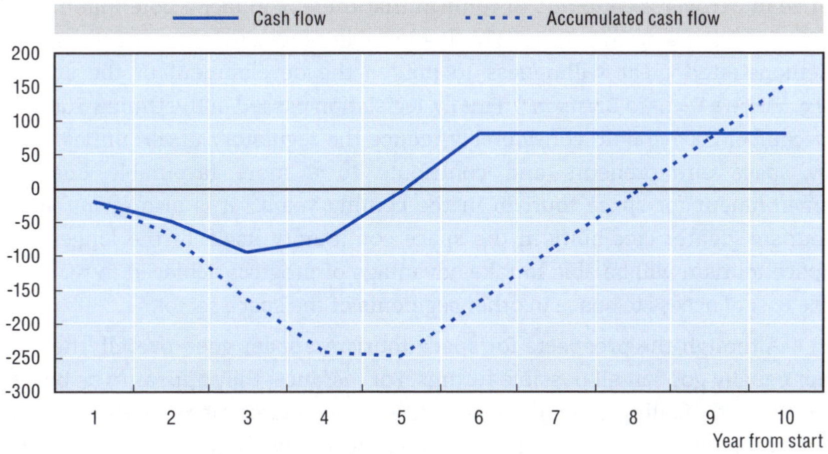

Source: Authors.

Model 2: Orbital tourism. The development of orbital tourism will depend on whether sufficient progress is achieved in reducing the cost of access to space in the coming decades. As noted, such progress could be achieved in two phases following the development of suborbital tourism in the coming years:

- *From 2015-20:* Advanced suborbital tourism, using technologies developed by defence and civilian R&D organisations for hypersonic propulsion and trans-atmospheric flight.
- *From 2025-30:* Development of orbital tourism, taking advantage of advanced space transport systems (probably two-stage-to-orbit systems, with the first reusable stage derived from the hypersonic trans-atmospheric vehicles introduced in the preceding decade).

However, so many uncertainties surround these potential phases of development that it is impossible to propose any real assumptions for a business model. In fact, as suggested in the Futron study and others, orbital tourism may well continue to develop as a mere continuation of the trend initiated by Tito's flight, *i.e.* space tourists simply taking advantage of available seats on ferry vehicles (Soyuz at the beginning, replaced by an improved ferry vehicle around 2009) servicing the ISS until new technology come on stream in the late 2020s.

Prospects for the future

A number of favourable factors should contribute to the development of space tourism in the coming decades. First, there is no technological barrier to the development of suborbital tourism; this should allow the development of the initial phase of the activity in the coming years. Second, the great interest generated by the X-Prize and the historic flight of SpaceShipOne, which won the prize in October 2004, tends to confirm that there is in fact a potential market for space tourism/adventure.[3] Third, a number of wealthy entrepreneurs have demonstrated great willingness to pursue the development of the activity (*e.g.* Virgin's Richard Branson[4]). Finally, legislation passed in the United States in December 2004 should considerably reduce the regulatory uncertainties faced by space entrepreneurs and contribute to a more favourable business environment for space tourism in the coming years.[5] It is also giving space tourism greater credibility in the space community itself. In the longer run, space tourism will be able to take advantage of progress achieved in reducing the cost of access to space in other segments of the space sector.

Although the prospects for space tourism appear good overall, they are not equally good in all possible futures. For instance, they appear to be best in the Smooth Sailing scenario where the open environment for commercial space should be favourable to new space ventures, while the more competitive climate should put strong pressures on costs. Moreover, technology transfers from the military should be easier. On the other hand, less public money may be devoted by the military to the development of hypersonic planes. In the Back to the Future scenario, business conditions are somewhat less favourable but could be compensated in part by the greater emphasis on military R&D for reducing the cost of access to space. There might, however, be some reluctance to transfer this technology to the commercial sector, in the light of its dual use nature and the fear that it might fall into the "wrong hands". Prospects are least encouraging in the Stormy Weather scenario, in which money is scarce, security concerns are very high on the policy agenda and restrict technology transfers from the military to the commercial sector, and discretionary expenditures for leisure and tourism are drastically cut.

Notes

1. In the NASA/STA study (1998), it is recommended that NASA demonstrate technology that would increase safety, reliability and comfort, and decrease unit costs all by factors of ten; learn how to deal with high rocket launch noise, atmospheric pollution and debris collision concerns; learn how to provide low-cost human habitation facilities in orbit; and learn how to remedy passengers' discomfort and space sickness.

2. Many "market studies" on space tourism have been conducted in the past, but they were based on very sketchy and limited polls. Hence, their usually very optimistic conclusions cannot be taken as a serious indication of market size.The only useful information that can be extracted from these studies is a rough estimate of the price elasticity of the demand curve, when taking into account, on the one hand, the price paid for Soyuz orbital flights to the ISS, and on the other, the estimated market response to much cheaper offerings, based on hypothetical advanced low-cost space transport systems. The most recent market study was conducted by Futron and Zogby International (Futron, 2003). It is based on a survey of 450 "qualified" individuals (i.e. annual income of USD 250 000 or a net worth of USD 1 million) who were surveyed in January 2002.

3. The Ansari X-Prize, announced in the mid-1990s, offered USD 10 million to the first private team to develop a reusable spacecraft capable of carrying humans to suborbital space twice in two weeks. It was won in October 2004 by Burt Rutan's SpaceShipOne, financed by millionaire Paul Allen.

4. Branson, the Virgin Group's founder and chairman, announced in autumn 2004 that he was licensing the SpaceShipOne technology from Allen's Mojave Aerospace Ventures to launch Virgin Galactic Airlines in the next few years. Virgin Galactic would offer suborbital spaceflights to passengers willing to pay USD 200 000 for the experience.

5. The Commercial Space Launch Amendments Act of 2004 (H.R. 5382) underwent many revisions after first passing the House of Representatives in March by a vote of 402-1. It was not until December 2004 that the bill was finally sent to the White House for the US President's signature.

Areas for Action

Introduction

Although the applications reviewed here are diverse and take place in very different contexts, there is a high degree of communality from a policy perspective with regard to the issues that are critical for their successful development. To facilitate the presentation and put them in a public policy context, these issues are organised into seven policy clusters:

1. Creation and preservation of a stable and predictable environment for the applications.
2. Maintaining a level playing field.
3. Equitable access to services.
4. Support to the effective generation, distribution and use of information.
5. Support to the development of standards, compatibility and interoperability.
6. Development of physical and institutional infrastructure.
7. Support to basic R&D.

Each cluster represents what might be called a "generic" policy area which cuts across several of the applications explored in the preceding sections. Each is backed up by supporting examples taken from specific applications. The clusters reflect different types of actions that governments may take to foster economic and social development.

1. Creation and preservation of a stable and predictable environment for space applications

An important pre-condition for a thriving industry is a reasonably stable business environment at national and international levels and the confidence of market participants that an open multilateral system of trade and investment will be maintained and fostered. An unstable and unpredictable public policy environment increases the level of uncertainty for entrepreneurs, affects adversely their investment decisions and forces them to take a short-term approach to their business activities.

The significance of this phenomenon is clearly evident in the case studies described in this study. For instance, the adverse impact of legal and regulatory uncertainties on business decisions was noted for satellite entertainment (e.g. uncertainties regarding the rules applying to spectrum allocation); Earth observation (e.g. uncertainties related to the application of "shutter control" regulations); location-based services (e.g. uncertainty regarding the legal consequences of a disruption of the signal both for the signal provider and the users of the signal); and space tourism (e.g. uncertainty regarding the legal regime applicable to space tourism planes).

Another important area concerns uncertainties relating to liability, notably for emerging applications. This applies to telehealth (e.g. liability attached to a tele-consultation); location-based services (liability linked to a failure of the navigation signal); and space tourism (liability of the space tourism entrepreneur vis-à-vis clients and third parties).

It is also clear that intellectual property issues are essential and cut across applications. Their importance is likely to increase as the private sector plays a growing role in the development of new space applications, in terms of both technical and financial contributions. Moreover, intellectual property protection is likely to be a key factor in the successful design and implementation of space business models involving public/private collaboration. Hence, an effective institutional and regulatory framework governing intellectual property will be needed to provide the kind of legal certainty on which space business can successfully thrive.

2. Maintaining a level playing field

The creation and maintenance of a level playing field is a major concern of economic policy in a market economy. If resources are to be used efficiently, all competitors should be on the same footing. Barriers to entry and exit should not be artificially raised. Notably, this means that existing rules should not give legacy actors an undue advantage and that regulations should be technology-neutral.

One lesson to emerge from the case studies is that creating and preserving a balanced competitive environment for space applications has three important (partly interrelated) dimensions: the degree to which the public sector is involved in investment in and operation of the application; the role of regulations that shape the economic environment; and the support provided by government to specific, sometimes competing, technologies.

The problem arises for instance regarding Earth observation (coexistence in the same market of public and private actors facing different sets of constraints); satellite entertainment (e.g. national ownership rules); and location-based services (e.g. competition between space-based and ground-based provision of navigation data).

3. Equitable access to services

"Equitable access" is a broad general policy principle in democratic societies. It is based on the premise that all citizens should be equal before the law and should have equal right of access to public services, as well as to private services that are deemed "essential", whenever it is economically feasible. This principle has been widely used, in the field of telecommunications for instance, to justify the cross-subsidy of rural subscribers by urban ones. In such cases, the application of the equitable principle involves a trade-off between equity and efficiency, and the extent to which lack of economic feasibility can be invoked in practice to justify exceptions to the application of this principle varies from country to country, depending on whether "individualistic" or "egalitarian" principles dominate.

For example, the issues surrounding equity of access to satellite-provided services extends beyond the "divide" between urban and remote areas to address questions of equal treatment of individuals and national entities. Indeed, the equitable access principle may also extend to access to knowledge and information derived from space activities in general. In view of Article I, paragraph 1 of the Outer Space Treaty, which stipulates that exploration and use of outer space "must be effected for the good and in the interest of all countries, regardless of their state of economic and scientific development, being the attribute of all mankind", it has been argued that the knowledge and information resulting from space activities should be available to all countries, and in particular to developing countries (Conseil Économique et Social, 2004).

4. Generation, distribution and use of information

Space-based services (i.e. space communications, Earth observation and location-based services) can be considered as information services. As such, they present characteristics that set them apart from other goods. Unlike standard economic goods, information can be exchanged and used many times; the consumption by one user does not diminish the amount available for other users. New information products can be easily created by forming different combinations of other information products, a process that is made easier by digitalisation.

These characteristics make the pricing of information particularly difficult or force producers of content to find ways other than direct pricing to finance production (*e.g.* advertising, product placement, sponsoring). It also justifies government intervention (*e.g.* public production or support to the production and distribution of certain forms of information which, although valuable for society as a whole, would not be produced and distributed otherwise), notably when the information has a strong social and cultural dimension. Government intervention is also justified to protect the privacy and confidentiality of personal information.

Regarding the pricing of data, issues arise for instance for EO data (public data are made available at no charge or at the marginal cost of reproduction, while private operators have to charge prices that cover their costs). Moreover, privacy and confidentiality issues affect the future of space applications such as telehealth (need for secure and confidential communications links); satellite entertainment (the greater scope for monitoring subscribers' viewing habits offered by new devices such as digital video recorders (DVRs) raises fear of misuse); location-based services (unsolicited wireless advertising could be an obstacle to their widespread adoption).

5. Standards, compatibility and interoperability

In most areas of economic activity, notably in high-technology sectors such as space, the development of open standards plays a major role in the market place by providing the basis for reaping economies of scale, for fostering the compatibility and interoperability of equipment, and for strengthening competition among equipment manufacturers.

Standards have a strong public good dimension. However, experience suggests that although governments should support their development, standards should not, as a general rule, be set by public fiat, and in particular governments should not try to use standards as a tool for protecting domestic industry (*e.g.* the setting of the colour TV standards). However, governments must ensure that the formulation of standards by particular groups of firms does not give them an undue advantage in the market place. Governments should therefore act both as coaches and as umpires to ensure that standards remain really open.

The case studies clearly demonstrate that greater compatibility between technological systems, standards, licensing practices and so on, are crucial to the future development of all the applications examined in this report. In the case of telehealth, for instance, licensing requirements are a major obstacle; in Earth observation, lack of data compatibility is a serious problem; in location-based services, global interoperability of services and applications is essential to the creation of seamless, wireless location bases services.

6. Development of physical and institutional infrastructure

Governments have the main responsibility for the development of infrastructure. However, they can often contract the development and operation of such infrastructure to the private sector. Moreover, the private sector may participate in the financing of infrastructure that can be used to generate private revenues. In such cases, the public authority may either pay directly a share of the cost of the infrastructure or guarantee a revenue stream that is sufficient to induce the private partner to finance the infrastructure entirely on its own.

A recurring issue in the case studies is the important role of infrastructure and the extent to which public authorities should be involved in its provision and operation. For instance, government may encourage the extension of broadband services to rural and remote areas by private satellite operators, although this may violate the technology neutrality of public policy. Location-based services are provided by an infrastructure that has a strong public good characteristic (*e.g.* provision of safety and security) but can also be used for commercial purposes. EO data can be produced either by public agencies or by private firms.

7. Support of basic R&D

R&D activities are a major source of economic growth today and will become increasingly so as the technological sophistication of our economies increases. Governments have a key role to play in this regard, notably for ensuring that the level of funding and the allocation of funds to various areas of basic research are adequate and reflect public policy objectives. Governments also need to ensure that research results are fully exploited by the private sector.

In some application fields where the benefits are largely of a "public good" kind, there would seem to be a fairly clear-cut case to be made for encouraging government support of R&D. This applies for instance to telehealth, location-based services, Earth observation for disaster management and space tourism, as the R&D ties in with the broader public interest in reducing the cost of access to space.

Conclusion

The preceding consideration of case studies has been a useful exercise from a policy perspective. It makes it possible to set the many policy issues raised by the development of the space sector into a concrete context. It is clear from the above analysis that, although each space application is different, some common problems arise, many of which are obviously the responsibility of governments. Moreover, the case studies clearly demonstrate that many of the issues with an important bearing on space applications fall outside the field of space policy *per se* and need to be considered in a much broader public policy context.

Bibliography

Telehealth and Entertainment via satellite

Analysys (2003), "Delivering the Broadband Home – New fixed and mobile services and devices: forecasts 2003 – 2008", Cambridge, December.

Beach, M. et al. (2001), "Evaluating Telemedicine in an Accident and Emergency Setting", *Computer Methods and Programs in Biomedicine*, Vol. 64 (3), pp. 215-223.

Bergman, D.(2001), "The Telehealth Interoperability Lab", Alberta Research Council, Alberta, November.

Bird, K. (1972), "Cardiopulmonary Frontiers: Quality Health Care via Interactive Television", *Chest* 61, pp. 204-205.

Conseil Économique et Social (2004), "La politique spatiale de recherche et de développement industriel, avis et rapports du Conseil Économique et Social", Report presented by Alain Pompidou, 23 June.

European Commission (2002), "eEurope 2005: An Information Society for All: An Action Plan to be Presented in View of the Sevilla European Council", 263 final, 21-22 June.

Hailey, D. et al. (2003), "Recent Studies on Assessment of Telemedicine: Systematic Review of Study Quality and Evidence of Benefits", Institute of Health Economics Working Paper 03-04, Edmonton, Alberta.

Hussein, Z. et al. (2001), "Assessment of Videoconferencing in Telehealth in Canada", Technology Report No. 14, Canadian Coordinating Office of Health Technology Assessment, Ottawa.

ICRC – International Federation of Red Cross and Red Crescent Societies (2001), *Communicators Guide*, Geneva.

IEG – Indian Express Group (2003), "Telemedicine: Emergence of the Virtual Doctor", Express Computer Online, Mumbai, India, 10 March.

Le Goff-Pronost, M. et al. (2003), "Planification de l'offre de soins et télémédecine", *Géographie, Économie, Société*, No. 5, pp. 359-378.

McIntosh, E. et al. (1997), "A framework for the economic evaluation of telemedicine", *Journal of Telemedicine and Telecare*, Vol. 3(3), pp. 132-139.

Medmarket Diligence (2003), "Telemedicine Market in State of Flux", Report, December.

Moore, M. (1993), "Elements of Success in Telemedicine Projects", Report of a research grant from AT&T, Graduate School of Library and Information Science, University of Texas at Austin.

Moore, M. (1999), "The Evolution of Telemedicine", *Future Generation Computer Systems*, Vol. 15, pp. 245-54.

Northern Sky Research (2002), "Broadband Satellite Markets 2002: A Comprehensive Source of Markets Forecasts and Industry Trends", Report, Cambridge Massachusetts, July.

Northern Sky Research (2003), "DVB-RCS and DOCSIS: Evaluating the prospects for standards-Based Broadband Satellite Growth", Report, Cambridge, Massachusetts, February.

PricewaterhouseCoopers (2004), *Global Entertainment and Media Outlook 2004-2008*, Fifth edition.

Reid, J. (1996), "A Telemedicine Primer: Understanding the Issues", *Innovative Medical Communications*, Billings, Montana.

Slide, A. (1991), *The Television Industry – A Historical Dictionary*, Greenwood Press, New York.

The Economist (2004), "The Health of Nations: A Survey of Health-care Finance", 17 July.

The Times of India (2004), "Education Satellite Launched in Karnataka", 29 January.

UNESCO (2002), *Open and Distance Learning: Trends, Policy And Strategy Considerations*, Paris.

United States (2004), *Innovation, Demand and Investment in Telehealth*, Office of Technology Policy, Department of Commerce, Washington, DC.

Winters, S.R. (1921), "Diagnosis by Wireless", *Scientific American*, No. 11.

Road traffic management via satellite

Angelides, J. (2003), "Choosing the Right Technology for Location-based Services", *Connect-World Asia-Pacific*, 22 January, www.connect-world.com, accessed 26 January 2004.

Bates, J. (2003), "Security Concerns Boosting Sales of GPS Devices", *Space News*, 30 June.

Blanchard, W. (2003), "Achieving GPS – Galileo Interoperability: The Challenges Ahead", *Space Policy*, Vol. 19, pp. 95-99.

DRAST – Direction de la Recherche et des Affaires Scientifiques et Techniques (2003), *Les applications des signaux satellitaires : exploration des usages envisageables à horizon 8-10 ans et rôle possible des pouvoirs publics*, Report prepared for the DRAST by Groupe CM International, October.

EC – European Commission (2003), "Business in Satellite Navigation: An Overview of Market Developments and Emerging Applications", March, Brussels.

EC (2004), "Inception Study to Support the Development of a Business Plan for the GALILEO Programme: Executive Summary Phase II", Brussels.

ESA – European Space Agency (2003), "Pay-as-you-go motoring Just Around the Corner", *ESA Press Release*, 9 September.

Foresight Vehicle (2004), *Foresight Vehicle Technology Roadmap, Technology and Research Directions for Future Road Vehicles*, February.

Frost and Sullivan (2004), "North American Passenger Vehicle Telematics and Remote Vehicle Diagnostics Markets", 1 April.

Garmin (2003), "Garmin Introduces GPS Receiver Targeting Automotive OEM Applications", *Garmin Press Release*, 18 November.

Grajski, K. and E. Kirk (2003), "Towards a Mobile Multimedia Age, Location-based Services: A Case Study", *Wireless Personal Communications*, Vol. 26(2-3), pp. 105-116.

IDC Research (2004), "US Cellular Location-Based Services 2004-2008 Forecast and Analysis: Mapping a Mobile Future", Global Information, Inc., 3 March.

IEA – International Energy Agency (2002), *Transportation Projections in OECD Countries*, OECD, Paris.

OECD (2001), *Towards Sustainable Transportation*, OECD Conference Proceedings, 9 October, OECD, Paris.

OECD (2003), *Road Safety: Impacts of new technologies*, OECD's Programme on Research on Road Transport and Intermodal Linkages (RTR), OECD, Paris.

OECD (2003b), *Trends in the Transport Sector 1970-2001*, European Conference of Ministers of Transport, OECD, Paris.

OMA – Open Mobile Alliance (2002), Location Interoperability Forum (LIF), "Interoperability Testing Group, The challenge with inter-operability in LCS", Document LIF201-v300.

PWC – PricewaterhouseCoopers (2001), "Inception Study to Support the Development of a Business Plan for the Galileo Programme: Executive Summary", TREN/B5/23-2001, PricewaterhouseCoopers.

Spinney, J. (2003), *A Brief History of LBS and How Open LS Fits Into the New Value Chain*, ESRI.

Stopher, P. (2003), "Reducing road congestion : a reality check", *Transport Policy*, September.

Telecompetition, Inc. (2003), *Worldwide Mobility Report 2003: Mobile Voice, Mobile Data, Subscriber, Revenue, and Penetration Rate Forecasts, 2002-2010*, August.

United Nations (2004), *World Urbanization Prospects: The 2003 Revision*, UN Population Division Report.

US DoT – Department of Transportation (2001), "Vulnerability Assessment of the Transportation Infrastructure Relying on the Global Positioning System", Report prepared for the DoT by the Volpe National Transportation Systems Center, 29 August.

Risk and disaster management

Baker, J.C., K. O'Connell and R. Williamson (2001), *Commercial Observation Satellites: At the Leading Edge of Global Transparency*, co-published by RAND and American Society for Photogrammetry and Remote Sensing.

Banks, R. (2002), "More sophisticated Models Aid Insurers", *Insurance Technology*, Summer.

BNSC – British National Space Centre (2001), *Market Sector Studies Programme – Insurance Market*, November.

CEOS – Committee on Earth Observation Satellites (2003a), *The Use of Earth Observing Satellites for Hazard Support: Assessments and Scenarios*, Final Report of the CEOS Disaster Management Support Group.

CEOS (2003b), "Improving Utilization of Earth Observation Satellite Data", Decisions of the 17th CEOS Plenary on Satellite Data Utilization, December.

CEOS (2004), *Newsletter No.22*, January.

CNES – Centre National d'Études Spatiales (2003), "Dossier: L'espace contre les risques naturels", *CNES Magazine*.

DRAST – Direction de la Recherche et des Affaires Scientifiques et Techniques (2003), "Les applications des signaux satellitaires : exploration des usages envisageables à horizon 8-10 ans et rôle possible des pouvoirs publics", Report prepared for the DRAST by Groupe CM International, October.

ESA – European Space Agency (2003a), "Signals from space enable earthquake detection", *ESA Press Release*, 31 October.

ESA (2003b), "Satellites assist planners preventing floods", *ESA Press Release*, 27 November.

FEMA (2003), FEMA *Disaster Costs, 1990 to 1999, www.fema.gov*, accessed 10 January 2004.

Frost and Sullivan (2003), *World Commercial Remote Sensing Imagery GIS Software Data and Value-added Services Markets*, July.

GEOSS – Global Earth Observation Satellite Systems (2004), *Draft GEOSS 10-Year Implementation Plan Technical Blueprint*, Washington, DC.

Harris, R. (2003), "Current policy issues in remote sensing: report by the International Policy Advisory Committee of ISPRS", *Space Policy*, Vol. 19, pp. 293-296.

Holt-Andersen, B. et al. (2004), Flood and Fire Pilot Study,,, ControlWare Report, Leuven, Belgium, September.

Iannotta, B. (2003), "Fire response officials feel impact of Landsat 7 glitch", *Space News*, 3 November.

International Charter "Space and Major Disasters" Secretariat (2003), *2nd Annual Report*, Executive Secretariat, May-December 2002, 25 June.

International Federation of Red Cross and Red Crescent Societies, *World Disaster Report 2002*, at *www.ifrc.org/PUBLICAT/wdr2002/index.asp*.

Mission Risques Naturels (2003), "Mission des sociétés d'assurance pour la connaissance et la prévention des risques naturels", at *www.mrn-gpsa.org/public/index.html*.

Mondello, C., G. Hepner and R. Williamson (2004), "10-Year Industry Forecast Phases I-III – Study Documentation", Prepared for the American Society for Photogrammetry and Remote Sensing (ASPRS), January.

Munich RE (2004), *Topics GEO – Annual Review of Natural Catastrophes 2003*, 25 February.

NASA (1998), "Integration of Remote Sensing and GIS with FEMA Flood Hazard Mapping", Report prepared for NASA by Sedona1 GeoServices, Inc., Series ARC-USU-002-97.

NASA-Joint Propulsion Laboratory (2002), "The Millennium Program – Earth Observing-1, General Presentation", Goddard Spaceflight Centre, August, *http://eo1.gsfc.nasa.gov*, accessed 3 February, 2004.

Nirupama, S.S. (2002), "Role of Remote Sensing in Disaster Management", Institute for Catastrophic Loss Reduction, ICLR Research Paper Series No. 21, The University of Western Ontario.

NOAA – National Oceanic and Atmospheric Administration (2004), *Economic Statistics for NOAA*, Third Edition, April.

OECD (2003a), *Emerging Risks in the 21st Century*, OECD, Paris.

OECD (2003b), Commercial Space Project Working Document, October.

REMSAT (2004), *European Space Agency REMSAT (Real-time Emergency Management via Satellite)*, *www.remsat.com*, accessed 11 February, 2004.

Roeser, H.P. (2003), "Cost Effective Earth Observation Missions, Fundamental Limits and Future Potentials", Institute of Space Systems, University of Stuttgart, paper presented at 4th IAA Symposium on Small Satellites for Earth Observation, Berlin, Germany, 7-11 April.

de Selding, P.B. (2003), "Spot Image Focuses on Serving Its Government, Military Customer Base", *Space News*, 18 November.

Swiss RE (2004), *Natural catastrophes and man-made disasters in 2003*, No. 1, 5 April.

UN ESCAP – United Nations Economic and Social Commission for Asia and the Pacific (2003), "Use of space technology applications for poverty alleviation: trends, strategies and policy frameworks, ESCAP works towards reducing poverty and managing globalization", Report ST/ESCAP/2309.

World Health Organisation (WHO) (2002), J. Adams and B. Wisner (ed.), *Environmental Health in Emergencies and Disasters: A Practical Guide*, World Health Organisation, Geneva.

WMO – World Meteorological Organisation (2002), "Economic framework for the provision of meteorological services", *WMO Bulletin*, Vol. 51(4), October, pp. 334-342.

WMO (2004), "Key Issues: Meteorological Satellites and the World Meteorological Organisation Programmes, Position Paper for the OECD Commercial Space Project", 24 February.

Space tourism

Abitzsch, S. and F. Eilingsfed (1992), "The Prospects for Space Tourism: Investigation on the Economic and Technological Feasibility of Commercial Passenger Transportation in Low Earth orbit", IAF Paper No. IAA-92-0155.

Collins, P. (2001), "The Prospects for Passenger Space Travel", Speech to the 4th Commercial Space Transportation Forecasting Conference, Arlington, Virginia USA, February 6-7.

Collins, P., R. Stockmans and M. Maita (1995), "Demand for Space Tourism in America and Japan, and Its Implications for Future Space Activities", *Sixth International Space Conference of Pacific-Basin Societies*, Marina del Rey, California, *Advances in the Astronautical Sciences*, Vol. 91, pp. 601-610.

Crouch, Geoffrey I. (2001), "The Market for Space Tourism: Early Indications", *Journal of Travel Research*, Vol. 40, November 2001, pp. 213-219.

Crouch, G.I. et al. (2004), "Space Tourism Through the 21st Century", *Tourism: State of the Art II Conference*, University of Strathclyde, Glasgow, 27-30 June.

Futron (2003a), *Space Tourism Market Study*, Bethesda, Maryland.

Futron (2003b), "The Viability of Human in Space: a Business Perspective", AIAA, Session 2-BPS-1, 2 September.

International Space University (2000), *Space Tourism: From Dream to Reality*, Summer Session Program 2000, Strasbourg, France.

NASA/Space Transportation Association (1998), *General Public Space Travel and Tourism*, NASA Report, Washington DC, March

Penn, J. and C. Lindley (1997), "Requirements and Approach for a Space Tourism Launch System", Paper IAF-97-IAA.1.2.08.

Sarigul-Klijn, M. and N. Sarigul-Klijn (2003), "Flight Mechanics of Manned Sub-Orbital Reusable Launch Vehicles with Recommendations for Launch and Recovery", AIAA 2003-0909.

Space Adventures (2004), *www.spaceadventures.com*.

US DoC – US Department of Commerce (2002), "Market Opportunities in Space: The Near-Term Roadmap", prepared by DFI International for US Department of Commerce Office of Space Commercialisation, December.

Wertz, James R. (2000), "Economic Model of Reusable Versus Expendable Launch Vehicles", *IAF Congress*, Rio de Janeiro, Brazil 2-6 October.

WTO – World Tourism Organisation (1999), *www.world-tourism.org*.

ANNEX B

The Space Project Steering Group

The Steering Group

At the beginning of the project entitled "The Commercialisation of Space and the Development of Space Infrastructure: The Role of Public and Private Actors" (the "Space Project"), a Steering Group was set up to provide overall advice to the OECD Project Team. It was composed of approximately 25 high-ranking experts and decision makers from public and private entities in the space and non-space sectors that contributed financially to the project. Throughout the duration of the project, there were four meetings of the Steering Group (April 2003, October 2003, May 2004 and December 2004).

Chairman

Michael OBORNE
Director of the OECD International Futures Programme
OECD

The members

Some members of the Steering Group were replaced during the two years of the Space Project and/or assisted by other experts from their organisations and/or third parties. The Steering Group members for Phases III, IV and V of the project were.

David ABELSON
Deputy Vice President International
Lockheed Martin Space Operations Company, United States

Josef ASCHBACHER
Programme Coordinator, Directorate of Earth Observation Programmes
European Space Agency

Walter AIGNER
Director, HiTec Marketing, Austria
(Adviser to Andrea Kleinsasser)

Eugène BERGER
Premier Conseiller de gouvernement, ministère de la Culture,
de l'Enseignement supérieur, et de la Recherche,
gouvernement du grand-duché de Luxembourg

Gianluca BETELLO
Senior Vice President, Studies, Planning and Strategic Control
Alenia Spazio SpA, Italy
(Replaced Giuseppe ARIDON and Luca TONINI)

Mara BROWNE
International Relations Specialist, NOAA Satellite and Information Services
US Department of Commerce, United States
(Replaced Timothy STRYKER)

Bill COWLEY
Director, Institute for Telecommunications Research (ITR)
University of South Australia, Australia

Pierre DECKER
Conseiller de gouvernement, ministère de la Culture, de l'Enseignement
et de la Recherche, Luxembourg

Christoph EBELL
CTI International, Program Manager CTI Asia Operations
EUREKA National Project Coordinator, Science/Industry International Relations
Federal Office for Professional Education and Technology OPET KTI/CTI, Switzerland
(Replaced Seta BOROYAN)

Patrick EYMAR
Vice President, Futures Projects, Launchers Directorate
EADS Space Transportation, France
(Replaced Ulrich BECK)

Paula FREEDMAN
Director of Space Applications and Transportation
British National Space Centre, United Kingdom

Roel R.R. HUIJSMAN-RUBINGH
Project Director
Ministry of Health, Welfare and Sport, The Netherlands

Chin-Young HWANG
Head, Department of Policy Studies and International Relations
Korea Aerospace Research Institute, Korea

Andrea KLEINSASSER
Manager, Space Research and International Technology Affairs
Federal Ministry for Transport, Innovation and Technology, Austria

Karl-Heinz KREUZBERG
Head of Strategy Department
Directorate of Strategy and External Relations, European Space Agency

Christine LEURQUIN
Senior Manager European Programmes
SES GLOBAL, Belgium

Johan LINDEMAN
Team Manager, Aeronautics and Space
Ministry of Economic Affairs, The Netherlands
(Replaced Leon NOORLANDER)

Leon NOORLANDER
Policy Advisor, Directorate-General for Innovation
Ministry of Economic Affairs, The Netherlands

Marcello RICOTILLI
Director, Space Program
Telespazio, Italy

Jostein RONNEBERG
Director of Application Development
Norwegian Space Centre, Norway

Timothy STRYKER
Chief, Satellite Activities Branch, International and Interagency Affairs
NOAA Satellite and Information Services
US Department of Commerce, United States

Per TEGNÉR
Chairman and Director General
Swedish National Space Board, Sweden

Didier VASSAUX
Délégué à la Stratégie et à la Prospective Spatiale
Centre National d'Études Spatiales (CNES), France

Frederik VON DEWALL
General Manager and Chief Economist
ING Group, The Netherlands

Monique WAGNER
Chef du département espace
Services fédéraux des affaires scientifiques, techniques et culturelles, Belgique

Jon WAKELING
Office of the BT Group Technology Officer
British Telecom, United Kingdom

During the first and second phases of the project, Canada was represented by Jean-Marc CHOUINARD (Canadian Space Agency) and Ian PRESS (Canadian Ministry of Natural Resources).

ANNEX C

Experts Involved in the Project

Contributing experts

Throughout the project, several experts were called upon to draft papers on various topics. These reports provided a very valuable factual and conceptual basis for the work conducted by the project team.

Raymond BOUCHARD (†)
Drachma Denarius Applied Futures Research and Strategic Planning
Ottawa, Canada

Alain DUPAS
International Consultant
Paris, France

Michel FOUQUIN
Deputy Director of the CEPII
Associate Professor at the University of Paris I
Paris, France

Henry R. HERTZFELD
Senior Research Staff Scientist
Space Policy Institute, Center for International Science and Technology Policy
George Washington University
Washington, DC, United States

Birgitte HOLT-ANDERSEN
ControlWare CVOH
Waterloo, Belgium

Thomas KANE
Department of Politics and International Studies
University of Hull, England

Molly MACAULEY
Senior Fellow, Energy and Natural Resources Division
Director, Academic Programs, Resources for the Future
Washington DC, United States

Matthew MOWTHORPE
Department of Politics and International Studies
University of Hull, England

Walter PEETERS
Dean, International Space University
Strasbourg, France

Frans G. VON DER DUNK
Director, International Institute of Air and Space Law
Leyden University, The Netherlands

International organisations consulted

Representatives of international organisations were invited to contribute their expertise in the course of the project. Some international organisations sent representatives to Project Steering Group meetings and/or contributed working papers. The representatives of those organisations are listed below.

Sergio CAMACHO
Director
United Nations Office for Outer Space Affairs (UN OOSA)
Vienna International Center

Mohamed ELAMIRI
Director
Air Transport Bureau
International Civil Aviation Organization (ICAO)

Michelle GAYER
Medical Officer
World Health Organisation (WHO)

Rodolfo de GUZMAN
Director, Strategic Planning Office
World Meteorological Organisation (WMO)

Yvon HENRI
Bureau des Radiocommunications
International Telecommunications Union (ITU)

Daniele GERUNDINO
Assistant Secretary-General, Strategies Management
International Organisation for Standardisation Standardization (ISO)

Tomoko MIYAMOTO
Senior Counsellor
Patent Law Section, Patent Policy Department
World Intellectual Property Organisation (WIPO)

Robert MISSOTTEN
Senior Programme Specialist, Earth Sciences Division
United Nations Educational, Scientific and Cultural Organization (UNESCO)

Martin J. STANFORD
Principal Research Officer
International Institute for the Unification of Private Law (UNIDROIT)

External experts/professionals consulted

A large number of space and non-space professionals contributed their views, through meetings and via e-mail, in the course of the project. The OECD project team is particularly grateful for their input. The following is a non-exhaustive list of the external experts/professionals consulted.

David ASHFORD (Bristol SpacePlanes, United Kingdom)

Rashid L. BASHSHUR (University of Michigan, United States)

Luiz BEVILACQUA (Brazilian Space Agency, Brazil)

Hélène BEN AIM (Ministry of Research and New Technologies, France)

Jean-Luc BESSIS (Centre National d'Études Spatiales, France)

Jerome BEQUIGNON (Ministère de l'Intérieur, France)

Yves BLANC (Eutelsat, France)

Mark BRENDER (Space Imaging, United States)

Brian CHASE (The Space Foundation, United States)

Jean-Marc CHOUINARD (Canadian Space Agency, Canada)

Patrick COLLINS (University of Azabu, Japan)

Geoffrey I. CROUCH (La Trobe University, Australia)

Lucien DESCHAMPS (Centre National de la Recherche Scientifique, France)

Patrick FRENCH (Northern Sky Research, United States)

Louis FRIEDMAN (The Planetary Society, United States)

Alain GAUBERT (Eurospace, France)

Marc GAUDRY (Université de Montreal, Canada)

Laurent GATHIER (Dassault, France)

Neil GOLBORNE (Department of Trade and Industry, United Kingdom)

Laurence GREEN (Ofcom, United Kingdom)

Michael HALES (NOAA, United States)
Ray HARRIS (University College London, United Kingdom)
Paul HEINERSCHEID (Satlynx, Luxembourg)
Mark HEMPSELL (University of Bristol, United Kingdom)
Holger ISCHEBECK (Eutelsat, France)
Ram JAKHU (Mac Gill University, Canada)
Steve JENNINGS (DigitalGlobe, United States)
Alexey KOROSTELEV (Russian Space Agency, Russia)
Wade LARSON (MacDonald, Dettwiler and Assoc. Ltd., Canada)
André LEBEAU (France)
Pierre LIONNET (Eurospace, France)
John LOGSDON (Space Policy Institute, United States)
Philippe MUNIER (Spot Image, France)
Charles MONDELLO (Pictometry, United States)
Christopher MYERS (Lockheed Martin, United States)
John MURTAGH (Infoterra Ltd., United Kingdom)
Gopalakrishnan NARAYANAN (Indian Space Research Organisation, India)
Xavier PASCO (Fondation pour la Recherche Stratégique, France)
Timothy PUCKORIUS (Orbimage, United States)
Philippe PUJES (Ministry of Research and New Technologies, France)
Isabelle SOURBÈS-VERGER (Fondation pour la Recherche Stratégique, France)
John SUNDQUIST (Lockheed Martin Navigation Systems, United States)
Phillipe TROYAS (EADS, France)
Edmund WILLIAMS (European Space Agency)
Oleg VENTSKOVSKY (Yuzhnoye State Design Office, Ukraine)
Yaroslav YATSKIV (National Space Agency of Ukraine)

OECD experts

Several OECD colleagues were called upon to provide comments on early drafts so as to ensure that the work conducted by the project team fully took into account related work in other parts of the Organisation.

Patrick DUBARLE (Public Governance and Territorial Development)
Jean GUINET (Directorate for Science, Technology and Industry)
Tom JONES (Environment Directorate)

Sandrine KERGROACH-CONNAN (Directorate for Science, Technology and Industry)

Sam PALTRIDGE (Directorate for Science, Technology and Industry)

Dirk PILAT (Directorate for Science, Technology and Industry)

Danny SCORPECCI (Directorate for Science, Technology and Industry)

Simon UPTON (OECD Round Table on Sustainable Development)

Dimitri YPSILANTI (Directorate for Science, Technology and Industry)

Acronyms

ACTS	Advanced Communications Technology Satellite
AGP	Agreement on Government Procurement
ASAT	Anti-satellite
ASI	Agenzia Spaziale Italiana
ATM	Air traffic management
ATV	Automated transfer vehicles
BBI	Broad-Band Interactive System
BIRD	Bi-spectral InfraRed Detection (DLR)
BNSC	British National Space Centre
CCSDS	Consultative Committee for Space Data Systems
CEOS	Committee on Earth Observation Satellites
CGMS	Co-ordination Group for Meteorological Satellites
CNES	Centre National d'Etudes Spatiales (France)
CNSA	China National Space Administration
CONAE	*Comisión Nacional de Actividades Espaciales*
COPUOS	Committee on the Peaceful Uses of Outer Space (UN)
COS	Commercial observation satellites
COSPAR	Committee on Space Research
CSA	Canadian Space Agency
CTBT	Comprehensive Nuclear Test Ban Treaty
CWAAS	Canadian Wide Area Augmentation System
DARPA	Defense Advanced Research Projects Agency (US)
DART	Demonstration for Autonomous Rendezvous Technology (US)
DBS	Direct broadcast satellites
DGA	*Délégation Générale pour l'Armement* (France)
DGPS	Differential GPS
DLR	German aerospace research centre (Deutsche Forschungsanstalt für Luft- und Raumfahrt)
DMSG	Disaster Management Support Group
DoC	Department of Commerce (US)
DoD	Department of Defense (US)
DTH	Direct-to-home
DTM	Digital terrain mapping

EBIT	Earnings before interest and taxes
ECSS	European Co-operation for Space Standardisation
EELV	Evolved Expendable Launch Vehicle
EGNOS	European Geostationary Navigation Overlay System
ELVs	Expendable launch vehicles
EO	Earth observation
EOS	Earth Observing System
ERS-1	European Remote Sensing Satellite
ESA	European Space Agency
ESF	European Science Foundation
ESTMP	European Space Technology Master Plan
EUROCAE	European Organisation for Civil Aviation Equipment
FAA	Federal Aviation Administration (US)
FAO	Food and Agriculture Organization
FCC	Federal Communications Commission (US)
FDI	Foreign direct investment
FFG	Austrian Research Promotion Agency (Österreichische Forschungsförderungsgesellschaft)
FTTP	Fibre-to-the-premises
GAGAN	India's GPS and GEO Augmented Navigation programme
GAO	Government Accountability Office (US)
GATS	General Agreement on Trade in Services
GATT	General Agreement on Tariffs and Trade
GEO	Group on Earth Observations
GHG	Greenhouse gas
GIS	Geographic information systems
GMES	European Global Monitoring for Environment and Security
GNSS	Global navigation space system
GOES	Geostationary satellite
GPS	Global Positioning System
GSO	Geostationary satellite orbit
HCV	Hypersonic cruise vehicle
HDTV	High definition television
IADC	Inter-Agency Space Debris Co-ordination Committee
IAOPA	International Council of Aircraft Owner and Pilot Associations
ICAO	International Civil Aviation Organization
ICT	Information and communications technologies
IEA	International Energy Agency
IGA	Intergovernmental agreement
IGOS	Integrated Global Observing Strategy

ILO	International Labour Organisation
IPO	Initial public stock offerings
ISA	International Space Agency
ISAS	Institute of Space and Aeronautical Science (Japan)
ISO	International Organization for Standardization
ISRO	Indian Space Research Organisation
ISS	International Space Station
ITAR	International Traffic in Arms Regulations (US)
ITS	Intelligent transport systems
ITU	International Telecommunications Union
JAXA	Japanese Aerospace Exploration Agency
JSF	Joint Strike fighter (US)
KARI	Korea Aerospace Research Institute
KSAT	Kongsberg Satellite Services
LAAS	Local Area Augmentation System (FAA US)
LBS	Location-based services
LEO	Low Earth orbit
LSA	Launch Services Alliance
MEXT	Ministry of Education, Culture, Sports, Science and Technology (Japan)
MODIS	Moderate Resolution Imaging Spectroradiometer
MSAS	Multifunctional Transport Satellite-based Augmentation System (Japan)
MSG	Meteosat Second Generation
MTCR	Missile Technology Control Regime
MTG	Meteosat Third Generation
NAFTA	North American Free Trade Agreement
NAL	National Aerospace Laboratory (Japan)
NASA	National Aeronautics and Space Administration (US)
NASDA	National Space Development Agency (Japan)
NGA	National Geospatial Intelligence Agency (US)
NIMA	National Imagery and Mapping Agency (US)
NIVR	Netherlands agency for aerospace programmes
NPOESS	National Polar-orbiting Operational Environment System
NPP	NPOESS Preparatory Project
NSAU	National Space Agency of Ukraine
OMB	Office of Management and Budget (US)
POES	Polar orbital satellites
PPP	Public-private partnership
PPS	Precise Positioning Service (GPS)
PRS	Public regulated service
PVR	Personal video recorders

ACRONYMS

RDM	Risk and disaster management
RLV	Reusable launch vehicles
ROSAVIAKOSMOS	Russian Aviation and Space Agency
SAI	Space Application Institute (Ispra, Italy)
SAR	Synthetic aperture radar
SBAS	Spaced-based augmentation systems
SIASGE	Sistema Italo-Argentino de Satélites para la Gestión de Emergencias
SLI	Space Launch Initiative
SPASEC	Panel of experts in the field of space and security (EU)
SPS	Standard Positioning Service (GPS)
TOMS	Total Ozone Mapping Spectrometer
TREES	Tropical Ecosystem Environment Observations by Satellites
TRIPS Agreement	Agreement on Trade Related Aspects of Intellectual Property Rights
TRMM	Tropical Rainfall Measuring Mission
UNESCO	United Nations Educational, Scientific and Cultural Organization
UNIDROIT	International Institute for the Unification of Private Law
VoIP	Voice-over-Internet Protocol
VSAT	Very small aperture terminal satellite
WAAS	Wide Area Augmentation System (US)
WBCSD	World Business Council for Sustainable Development
WHO	World Health Organization
WINDS	Wideband Internetworking Engineering Test and Demonstration Satellite (Japan)
WIPO	World Intellectual Property Organisation
WME	Weapons of mass effects
WMO	World Meteorological Organization
WTO	World Trade Organisation
WTTC	World Travel and Tourism Council

OECD PUBLICATIONS, 2, rue André-Pascal, 75775 PARIS CEDEX 16
PRINTED IN FRANCE
(03 2005 01 1 P) ISBN 92-64-00832-2 – No. 53925 2005

ERAU-PRESCOTT LIBRARY